Archaeological Practice as Politics and Ethics

John Carman

Archaeological Practice as Politics and Ethics

Springer

John Carman
International Centre for Cultural Heritage
University of Birmingham
Birmingham, UK

ISBN 978-3-031-69827-9 ISBN 978-3-031-69828-6 (eBook)
https://doi.org/10.1007/978-3-031-69828-6

This Springer imprint is published by the registered company Springer Nature Switzerland AG
The registered company address is: Gewerbestrasse 11, 6330 Cham, Switzerland

For Patricia, as always

Preface

This book has had a long gestation period. The idea first emerged at a student-organised conference on *Histories of Archaeology* held in Cambridge in 1995 where it was noted that while many had been interested in the history of archaeological ideas (reflected especially by the work of Bruce Trigger [2006]), there was much less focus on the history of archaeological practices. From then the idea lay dormant, although occasional publications of mine addressed aspects of practice over the years. In 2015 my own work on *Archaeological Resource Management* as a global practice was published with Cambridge University Press and the CUP editor suggested some kind of follow-up work that would develop two introductory chapters in that book, outlining how ARM had developed over the previous centuries to the present. This had appeal, but of course there is more to archaeology than ARM and any discussion of ARM practices could not dismiss them. As a result the scope of any such work would need to go beyond the confines of ARM.

Accordingly, this book explores archaeology as a discipline and a field of endeavour to understand the contribution it makes to the modern world. There is less concern with historical trajectories than was originally conceived but every chapter refers to the past of the discipline to examine how it has come to be today, and as such it is not (like so many histories of the field) a simple chronological litany of great practitioners and important innovations (e.g. Daniel 1981). In general it also avoids a direct concern with theoretical developments (e.g. Trigger 2006) while acknowledging the necessary connection between theory and practice. It is instead a history in the sense of a 'natural history', focusing on the range of different things that archaeologists do and how they do and have done it in order to explore the field as a distinctive aspect of social practice. The focus throughout is on *practice*, although quite a lot of practice inevitably involves thinking, theorising and indeed philosophising. It also concerns the structures—social, political, economic, professional, philosophical—in which archaeology is performed and the consequences for the discipline and its practitioners. The book is offered to those outside the field as a justification for archaeology—a field coming under increasing pressure from those who consider it (as so many humanities and social science disciplines) as an expensive luxury and who require immediate returns in narrow economic terms—and also as reassurance to fellow archaeologists facing such threats to their livelihoods and professional standing that what we do is indeed valuable.

In avoiding a strictly chronological evolutionary trajectory, the book is structured around themes rather than periods and both asks and answers questions about the perception of archaeology from both within and outside the field. In focusing on the 'what' and 'how' of archaeological practices of all kinds it necessarily engages with different ideas of what archaeology and archaeologists ought to be and the appropriate behaviours for the field. It locates archaeology in its wider contexts and looks beyond the field itself to those with whom archaeologists necessarily engage—which is everybody in the past, the present and the future. Reflecting modern online publication practice—and especially, as requested by the publisher, to accommodate those who read individual chapters rather than the whole text—the book is structured as a set of essays on aspects of archaeological practice. Those who choose to read the whole book in the sequence in which it is offered—to whom I offer thanks and highly commend for their perspicacity—will see that certain themes and ideas reappear throughout the book in different chapters, risking the accusation of repetition. In defence of this, I would emphasise that such themes relate to more than one aspect of practice and to ignore them would lead to the equally heinous crime of overlooking them. I therefore ask for such readers' forbearance.

The book draws widely on previous work by the author and others—looking at the techniques of the field, the people involved, the political and social impacts both on and by the field. In doing so it seeks to bring these different perspectives together to create a hopefully more holistic idea of what archaeology is, has been, and could be. It will be evident that I write from the perspective of an academic commentator on archaeology rather than a dedicated fieldworker or project lead (although I have of course paid my dues in fieldwork). The book therefore follows and builds upon previous work—especially my *Valuing Ancient Things* (1996), *Archaeology and Heritage* (2002) and *Archaeological Resource Management* (2015) as well as the co-authored *Archaeological Practice in Great Britain* (Schofield et al. 2011), the co-edited *Managing Archaeology* (Cooper et al. 1995) and *Oxford Handbook of Public Archaeology* (Skeates et al. 2012), and other shorter works. Although I am based in Britain and the book is written from within the Western world, inevitably reflecting this positionality, it is not from an entirely Anglo-centric perspective and recognises throughout that archaeology is a global phenomenon, reflecting the coverage of the journal *Archaeologies* of which I am an Editor. I make no claims to particular expertise in excavation, survey or other practical techniques and the book is in no sense a training manual or other guide to how to undertake archaeological projects. I do intend however that it will serve as a useful addition to the ammunition available to those seeking the resources to undertake their work and as an eye-opener for those who are less acquainted with the many roles archaeologists perform and the contribution archaeology makes to the wider world. I hope it succeeds in that aim.

Birmingham, UK
July 2024

John Carman

References

Carman, J. 1996. *Valuing Ancient Things: Archaeology and Law*. Leicester: Leicester University Press.

Carman, J. 2002. *Archaeology and Heritage: An Introduction*. London: Continuum Press.

Carman, J. 2015. *Archaeological Resource Management: An International Perspective*. Cambridge: Cambridge University Press.

Cooper, M.A., A. Firth, J. Carman, and D. Wheatley, eds. 1995. *Managing Archaeology*. London: Routledge.

Daniel, G. 1981. *A Short History of Archaeology*. London: Thames and Hudson.

Schofield, J., J. Carman, and P. Belford. 2011. *Archaeological Practice in Great Britain: A Heritage Handbook*, World Archaeological Congress Heritage Manual Series. New York: Springer.

Skeates, R., C. McDavid, and J. Carman, eds. 2012. *The Oxford Handbook of Public Archaeology*, Handbooks in Archaeology. Oxford: Oxford University Press.

Trigger, B.G. 2006. *A History of Archaeological Thought*, 2nd ed. Cambridge: Cambridge University Press.

Acknowledgements

Thanks are due to all those with whom I have worked, argued and generally interacted during my career in the field of archaeology. I am conscious of how many of them there are, and also that lists of names are always invidious. Inevitably, someone will expect to see their name and not find it, while others less worthy to be noted are indeed shown—and so rather than be guilty of such sins of omission and commission I leave names blank: but be assured that if you would expect to see your name listed here, it would be!

Thanks are due too to several generations of students—undergraduate, postgraduate and at PhD level—who (although none ever believe it) also helped mould my understanding and appreciation of archaeology.

I am grateful to staff at SpringerNature who were keen to see the work published and were kind to its author: especially Christi Lue who I met briefly at the Springer stall at conferences and Malini Arumugam who have seen the work from inception to production. I must thank too the two anonymous reviewers of the text who provided insightful commentary and critique. I am also thankful to MarySue Daoud, our SpringerNature editor for *Archaeologies,* for her constant help and advice, and the very pleasant encounter with Chris Wilby in Budapest.

Finally, the largest share of gratitude goes to Patricia Carman, my partner in life and work, manager, accountant, funder and (when I need it) goad, who has supported me through good times and bad, and has put up with those periods of obsession when writing dominated everything and those worse periods when writing needed to be done but I was unable to progress it. To her I owe more than I can ever repay, which is no excuse for not trying. I hope this work goes some small way to confirming her faith in me.

Birmingham, July 2024 John Carman

Contents

Abbreviations

AAA	American Association of Archaeologists (USA)
AACA	Australian Association of Consulting Archaeologists (Australia)
AHD	Authorised Heritage Discourse
AHM	Archaeological Heritage Management
AIA	American Institute of Archaeology (USA)
ARM	Archaeological Resource Management
BCE	Before the Common Era (of dating = BC)
CE	Common Era (of dating = AD)
CHM	Cultural Heritage Management
CIfA	Chartered Institute for Archaeology (UK)
CoE	Council of Europe
CRM	Cultural Resource Management
DGUF	Deutsche Gesellschaft für Ur- und Frühgeschichte e.V. (Germany)
EAA	European Association of Archaeologists (Czech Republic)
ECA	Early Career Archaeologist
ECR	Early Career Researcher
EMI	Electromagnetic Induction
EU	European Union
GPR	Ground Penetrating Radar
GPS	Global Positioning System
ICOMOS	International Committee on Monuments and Sites (France)
IIA	Institute of Irish Archaeologists (Ireland)
ISIS	Islamic State in Syria (also known as Daesh)
IUPPS	International Union of Pre- and Protohistoric Societies
KCPIRIA	Korea Cultural Properties Investigation and Research Institute Association (Korea)
PA	Public Archaeology
PEF	Palestine Exploration Fund (UK)
RPA	Register of Professional Archaeologists (USA)
SAA	Society for American Archaeology (USA)
SHA	Society for Historical Archaeology (USA)

SOPA	Society of Professional Archaeologists (USA)
UK	United Kingdom
UNESCO	United Nations Educational, Scientific and Cultural Organisation
US(A)	United States of America
WAC	World Archaeological Congress (USA)

List of Tables

Chapter 1
Introduction: Archaeology as Performance

The term 'archaeology' has multiple uses—it describes at once what people called 'archaeologists' do, the materials they retrieve from their activities and on which they work, and is also used more metaphorically in the sense of delving into any kind of past, especially one that may be arcane or deemed to be pointless. In this book the term is limited to refer to the work that archaeologists do. In saying that archaeology is 'performed' is not to deny the reality of that practice—it is in no sense any kind of pretence, as a performance in a play may be (although see passing reference below; and Pearson and Thomas 1994)—but it also catches the range of accepted practices that fall within the field that is archaeology. Not everything that an archaeologist does is archaeology—believe it or not, we are all normal human beings with families and friends and other interests and sometimes other employment as well—but when we act as archaeologists we do certain kinds of things and only certain kinds of things. That range of things though is wide and various and it is the purpose of this book—or one of them—to explore that wide range of practices and what they signify.

Distinguishing what one does as an archaeologist from other things one may do is however sometimes very difficult if not impossible. For those for whom archaeology is not just about an attempt to study the past as objectively as possible but is also an active engagement in the present and future, the roles of social or political activist can merge with that of archaeologist. Even for those who attempt to retain a separation between their professional and personal social existence, others may see much of what they do as inevitably politically and ideologically driven. So in doing archaeology one is also a social actor and necessarily engaged with other aspects of the world in which we live and the other people who inhabit it. It may however be possible to identify practices that are essentially archaeological from others that are not—and to distinguish the specifics of the role of 'archaeologist' from that of citizen or indeed human being. This is what this book will attempt to do—although taking the widest possible perspective on what it is possible to do under the label

J. Carman, *Archaeological Practice as Politics and Ethics*,
https://doi.org/10.1007/978-3-031-69828-6_1

'archaeologist'. To start, this chapter will consider archaeology as a realm of practices and how they are variously considered by their practitioners.

Archaeology as Work

Others have taken particular note of archaeology as an arena of action and that is what is meant by 'work' in this context. Of course archaeology is also work in the narrow sense of an area of professional practice for which individuals are paid. Archaeologists work as archaeologists in a surprisingly wide range of contexts across the globe—in academic institutions, in state bureaucracies, in museums and other cultural organisations, in private corporations, in charities, and for international non-governmental organisations, as well as in self-employed roles. Others trained and educated as archaeologists may work in other roles—as lawyers, accountants, bankers, diplomats, soldiers, educators, technicians, plumbers or electricians—and undertake their archaeology as a hobby or volunteer activity. Yet others with no formal training or education in archaeology may also conduct archaeology in groups or alone. It will depend upon the rules that apply in a specific context as to who may rightly declare themselves to be archaeologists officially, but for the purpose of this work anyone who does any of the kinds of activities considered to be 'archaeological' will be considered as an archaeologist because our focus here is specifically on what archaeologists *do* and anyone who does these things—whatever other status they may hold—will be included as a member of the community of archaeologists. It follows of course that those who may do things that are similar to what archaeologists do but do it differently—in a manner that is not 'archaeological' whatever that may be determined to mean—are not to be classed as archaeologists. One of our purposes will be to establish such distinctions and their legitimacy in different contexts. This may not prove an easy task!

Everill (2009) has reviewed the development of a professional archaeology in the UK since the nineteenth century, demonstrating the falsity of the rhetoric of the field that professionalism (in the narrow sense) is a recent development: he is able to show that from the earliest emergence of the field skilled individuals were employed to conduct fieldwork; although their contribution may not have been readily acknowledged either at the time nor in later histories of the field, hence his designation of them as *Invisible Diggers*. Periodic studies of the status of professional archaeology in various territories have augmented our understanding of the profession as it exists today: Flatman (2011, 35) lists several from across the globe from the early 2000s and this has continued since (see e.g. https://profilingtheprofession.org.uk/ for the UK; Aitchison 2010 for Europe; ARI 2006 for the USA). Such professional archaeologists tend to be young, well educated (to degree level or higher) and relatively poorly paid, often also with poor prospects due to lack of a clear career structure. They are also increasingly likely to be women, at least in Europe, and further disadvantaged by gender pay gaps (Lazar et al. 2014).

A trend over recent decades—alongside but somewhat paradoxical given the generally low status of a career in fieldwork—has also been the decline of opportunities for amateur and avocational work in archaeology: as requirements that archaeology should be performed by qualified professionals have become more widespread, especially in Europe following the adoption of the so-called Valletta Convention (Council of Europe 1992), amateurs have found themselves increasingly excluded from the ability to investigate sites. This is especially because of the requirement of Article 3ii of the Convention that "excavations and other potentially destructive techniques are carried out only by qualified, specially authorised persons" which is widely interpreted to mean only those professionally qualified and accredited by relevant organisations: for those outside the Anglophone world it means most frequently accreditation by the State. The rise of so-called Preventive Archaeology across the world (Naffé et al. 2008; Hermon and Vassallo 2012) has also encouraged the use of professionals in advance of development work, further excluding amateurs from the opportunity to investigate sites. Of course, excavation by amateurs is in practice no different from excavation by professionals so long as it is to a similar standard, but it is not paid and it is usually impossible to assert the qualified status of the amateur to justify payment by commercial or other clients.

Taking a less economic view of archaeological work, Edgeworth (2006a) and his contributors have examined aspects of archaeological practice from an ethnographic perspective, locating us in terms of a particular culture. It is also very much a focus on how archaeological practice produces archaeological objects as discussed in his earlier work (Edgeworth 2003) as well as how different points of view of a site result in quite different understandings of the site itself. Edgeworth (2006b, 9–14) notes two parallel trends in ethnographic approaches to archaeology: one is a focus on how archaeologists relate to and treat their material (e.g. Lucas 2001; Chapman 2000); and a second is a concern with how archaeologists relate to other human beings with an interest in the site, either as a form of stakeholder (although they may not use the term itself) such as a local or descendent community (see e.g. Marshall 2002), or as visitors to sites (McClanahan 2006), or indeed as other groups of archaeologists (Holtorf 2006). Firmly in the tradition of examining fieldwork as an interpretive practice, Cobb et al. (2012) focus on the relationship between practice and archaeological theory by an emphasis on the *site* where the fieldwork is conducted as the place where interpretation is produced (see also Hodder 1999). This reflects the earlier work of Edgeworth (1990, 2003) who emphasises how it is specifically the activity of the archaeologist that produces the archaeological object, rather than it existing independently of the archaeologist.

In somewhat similar vein, Shanks and McGuire (1996) have described archaeology specifically as a craft in which particular skills and knowledges are exercised. As they put it, "Archaeologists take what's materially left of the past and work on it intellectually and physically to produce knowledge" (Shanks and McGuire 1996, 76), so for them the *work* of archaeology is the production of knowledge about the past, using the material that is left to us from pasts as a start-point to purposefully, practically and with expression actively design a model of the past that is viable. A later chapter of this book will consider this productive aspect of archaeological practice in more depth, but here it is the fact that archaeology is seen by its practitioners specifically as a process that *produces* which is the significant point. It means that archaeologists see themselves as makers of some kind: rather than being mere collectors of the residues of past lives (Salerno et al. 2019) or inconvenient parasites on commercial development projects (Cooper 2008), they actively contribute something that is new to the wider world. For Shanks and McGuire (1996) that contribution is new knowledge of pasts; for others it is a distinctive way of knowing the world through the direct encounter with the material (e.g. Hamilakis 2013), or a way to examine important social or political issues in the present (e.g. Sabloff 2008).

Tilley (1989) has specifically referred to excavation as a form of theatre, although he uses the analogy to argue for a reduction in the number of excavations to focus on the stores of already-excavated material worthy of examination. Using the idea of theatre in a more metaphorical manner, Pearson and Thomas (1994) engaged in a dialogue to identify to what extent the archaeological production of knowledge could be considered similar to a theatrical performance. Although similar to my use of 'performance' above it is not identical: in their work the affinity between a dramatic work and the interpretation of material remains from the past was directly addressed by reference to specific concepts that apply in both contexts: that of site, context, culture; and in relation to the production of meaning, that of creating a specific output—a narrative.

The work of archaeology is understood differently by different archaeologists in different contexts and who undertake archaeology for diverse reasons and with varying ends in view. It follows that there is also a range of understandings of the idea of archaeology itself and in turn who can be considered as archaeologist. This is the topic of the following sections.

Defining Archaeology

It is premature to give a definition of archaeology in this first chapter because the purpose of the book as a whole is, among other things, to achieve just this if indeed it is possible. However, it is possible to review the definitions and understandings of archaeology that others have provided.

First, the word itself. It derives from two Greek words—*arche* meaning 'ancient' and *logos* meaning 'word'. Combining the two in 'archaeology' translates either as 'ancient words' or 'words about the ancient'. In either case, the emphasis is upon words rather than things, which is somewhat at odds with more conventional ideas of what archaeology is (see below and throughout the book). Nonetheless, the first usages of the term did in fact relate to the study of ancient (especially classical Greek and Roman) writings, including inscriptions, and the term 'monument' was also used at the same time to refer to ancient texts (Carman 1996, 51; citing Newton 1851). It was during the 1850s that—in Britain at least—the focus of organisations such as the London-based Archaeological Institute and the Society of Antiquaries of London shifted from a broad concern with anything from the past to a specific interest in material culture, including built monuments and material retrieved from buried contexts. The latter are what we generally mean today when we talk of archaeology in the sense of the materials with which archaeologists engage. Of course, there had been an interest in monuments and artefacts from the past from much earlier (see also below on ethics), generally classed as 'antiquarian' rather than archaeological (see e.g. Levine 1986; Schnapp 1996; Schnapp and Kristiansen 1999), although many of the practices of antiquarians would be incorporated in the field of archaeology as it developed into the discipline we know today (Schnapp 1996; Schnapp and Kristiansen 1999; Carman 2015, 3–27).

The movement from antiquarianism to archaeology as understood today—where archaeology may stand classed with 'science' whether natural or social—took place in the later nineteenth century in Europe, where archaeology was part of the wider processes which also produced the nation state and European colonial empires. The adoption of an evolutionary framework for understanding human culture allowed an acceptance of Europeans' right to rule over those deemed lower on the developmental scale and indeed to appropriate the material evidences of the pasts of others to fill museums (Diaz-Andreu 2007; and see below on ethics). Archaeology was therefore more than a mere exercise based upon an interest in the past: it was a tool of an emergent and indeed dominant politics which declared certain kinds of behaviours as acceptable (and indeed 'right') and gave its practitioners an identifiable status which has not yet been lost (see also Smith 2004, 2006).

Today there seems to be quite widespread acceptance of a common and politically neutral definition of twentieth century archaeology as "the [scientific] study of past cultures through the analysis of surviving material remains" (quoted by Flatman 2011, 8). The popular vision of archaeology as "digging up the past"—with its emphasis on the techniques of excavation as an end in itself—was established relatively early in the last century, especially in publications by senior figures in the field for the mass reading audience (e.g. Woolley 1937; Wheeler 1955) although others such as Childe had sought to dispel the idea in their own works in the same period, which emphasised the interpretive products of archaeological work concerning the long-term developments in human history (Childe 1942, 1951).

Writing in the late 1980s, Binford, drawing upon inspiration from one of his teachers (Spaulding 1960) and his own work over several decades, claimed that archaeology is defined jointly in term of its subject matter and the manner of that study (Binford 1989, 3). He went on to assert that the subject matter of archaeology is "simply artifacts" rather than alternatives such as "human behaviour, … symbolic codes, … social systems, … ancient cultures, … ancient settlements, [or] the past", each of which he ascribed to other commentators (Binford 1989, 3). However, he also went on to assert that archaeology does not end with the study of artifacts and their attributes: from this, archaeologists go on to make inferences about the purpose of the artifact and how it relates to activity in the past. The key question for any archaeologist, he argues, is how to justify the inferences drawn from the study of artefacts, and the chapter goes on to show the weaknesses inherent in the work of the others he critiques. The particular set of approaches to the study of the past

Binford advocates are the topic of the remainder of his book, in which he offers archaeology as a form of anthropological science.

Binford, although influential in his time and followed in his approach by others still, was then and is subject to challenge by alternative views of archaeology. Although also defining archaeology in terms of artefact analysis—as "the discipline concerned with the recovery, systematic description and study of antiquities"—Clarke also asserted archaeology's independence from other fields including both history and anthropology:

> archaeology is archaeology is archaeology.... Archaeology is a discipline in its own right, concerned with archaeological data which it clusters in archaeological entities displaying certain archaeological processes and studied in terms of archaeological aims, concepts and procedures (Clarke 1968, 12–13).

While Hodder (1986, 176–177) generally accepted Clarke's notion of archaeology's separation from other fields, he went on to affirm its status as a discipline "linked to and contributing to general social theory" (Hodder 1986, 177) which is a theme since developed by himself and others. The essentially political nature of the study of the past has been brought to the fore by writers such as Shanks and Tilley (1987), Gathercole and Lowenthal (1990) and Hamilakis and Duke (2007). These three examples are all quite different in their approaches to the issues they raise but between them they emphasise how archaeology as a field cannot be considered as an objective study of abstract phenomena nor divorced from other concerns. For Shanks and Tilley (1987), the issues are those of the apparent 'objectivity' of archaeology as a science and to replace it with an ethical approach to studying humanity. For Gathercole and Lowenthal (1990) the issues are those of national appropriations of pasts to serve contemporary political interests which may not be those of archaeologists. For Hamilakis and Duke (2007) the issues are those of archaeology's relation to a dominating capitalist ideology with implications for how archaeologists engage with others and issues such as property.

All such works, in discussing issues of power and authority, emphasise the ethical responsibilities of archaeologists, taking the notion of ethics beyond a mere concern for 'proper' professional performance into the relations of archaeology with wider social and economic phenomena, but rather than subscribing to a convenient attitude that puts archaeology in a hierarchical relationship with others as in the nineteenth century, seeks to define a progressive and indeed radical role for archaeology in society and to challenge us to effect change in the world from below. These are themes that will necessarily re-emerge later in this book and indeed in sections below.

There we shall leave the attempt to define the field, but it is one we shall return to in the book's concluding chapter. What is nonetheless clear is that while there is widespread agreement—at least within the field—that archaeology is about using material from the past to make statements about that past, there is significant disagreement about what those statements should be. Wherever one stands, it is increasingly difficult to accept that archaeology represents an 'objective' approach to studying the past: even merely to claim to describe in a disinterested manner what a particular past was like, is itself to take a particular political and ethical stance.

Defining an Archaeologist

It is as inappropriate to define what we mean by 'archaeologist' here as to define the field. At its simplest, an archaeologist is anyone who does archaeology. It is of course much more complex than that because as seen in the previous sections, there is disagreement about what archaeology itself is and what it is for. In addition, first, as we shall see through the course of this book, and have seen above, archaeologists do not do just one thing: they perform many different things and work in many different contexts. Second, there is the issue of who is entitled to be described as an archaeologist.

According to Clarke (1968, 12), following his definition of the field (see above), an archaeologist is "the man [*sic*] who is responsible for" recovering and systematically studying antiquities. The usages are interesting: firstly and most obviously the use of the adult masculine noun which excludes all women and children from the title of archaeologist, intentionally or not; but second the emphasis on *responsibility* for the activity, which also excludes anyone acting in a purely technical or subordinate role to retrieve material. The only true archaeologist is therefore an adult male who takes responsibility for, and therefore asserts authority to control, the archaeological process. It can of course be argued (and may well be true) that Clarke never intended his words to imply such a narrow interpretation, but elsewhere (Carman 2015, 178–180) I have also written on the tendency of archaeologists as a collective to define others who may not be considered as archaeologists (and see Chap. 7 for more on this).

It can be argued, and I am tempted to, that there is no definition of who is an archaeologist other than by defining who is *not*; and as a breed we are quite keen to distinguish ourselves from others. An example is Flatman (2011, 38–41) who marks for disapprobation in particular:

- adherents to 'fringe' ideas about the past—those who believe in and especially peddle tales of ancient astronauts, Atlantis, or myths of inevitable racial or other superiorities based on dubious interpretations of archaeological findings;
- those who retrieve material from sites illegally or illicitly;
- those who damage sites and monuments, either illegally or sometimes with official State sanction;
- those who retrieve material from sites legally but not to add to human knowledge, and most frequently for commercial gain.

Flatman (2011, 41–43) argues that the reason these lie outside the scope of archaeology is because they fail to comply with some basic principles that apply to archaeological work, as represented in a series of international charters and codes of practice (of which more below), including:

- maintaining, developing, and promoting the highest standards of professional practice, including full transparency about their activities;
- actively discouraging and combating the trade in illicit antiquities;
- working to preserve the scientific integrity of a total site, which means minimising loss of features and material and ensuring proper recording for dissemination;
- recognising the rights of communities to control access to and information on their cultural heritage.

Flatman's concern (as implied in the subtitle of his book) is for the training of those who wish to work in archaeology in the narrow sense used in the opening to this chapter—as a paid professional in one of the sub-sectors of the field, rather than as an amateur or avocational researcher. As suggested in the opening to this chapter, and will be addressed further in this book, these sub-sectors are many: including for Flatman (2011) branches of government, national and local, museums and similar organisations, academia, development archaeology, voluntary community agencies, and as consultants. In both development archaeology and consultancy there is scope for freelance activity but also for employment in larger organisations, not all of them dedicated to archaeology, such as utility providers (of water, power, etc.) or architectural or construction services. Consultancy is not limited to commercial organisations: this too can be provided to communities in need of advice about their life is being affected by government or commercial activity of which they have no part. As indicated above, the issue of ethics arises since whether an archaeologist chooses to work with and for government and commerce, or for those who may be affected by such work, may well depend upon personal moral choices and political ideology.

To be an archaeologist does not necessarily imply a particular perspective on the world: we are as divided as any other community about issues of rights and wrongs, who should rule and who not, how that rule should be exercised, if at all, and to whom or what we owe duties of responsibility and care. It may be fair to say that in general archaeologists are more likely to subscribe to liberal politics rather than anything more authoritarian, although this also depends on the context from which they come: authoritarian control is the norm in some parts of the world and among some communities who also provide archaeologists. Otherwise, globally archaeologists are very commonly highly educated, to at least university degree level, as likely to be women as men, may work mainly in their own country or overseas, and may be young, middle-aged or elderly. The latter two factors may help to determine how well travelled an archaeologist may be: those who work away from their own country will certainly be well acquainted with other countries, and younger archaeologists may be more mobile generally; on the other hand the fact that major archaeological organisations -- such as the International Union of Pre- and Proto-historic Societies (IUPPS), the World Archaeological Congress (WAC), and the European Association of Archaeologists (EAA)—hold their meetings in different territories on each occasion also encourages travel across the globe. The majority of archaeologists can however be expected to come from the richer countries of the global

North, where archaeology has a longer and better-established pedigree, and archaeologists may have access to greater financial resources, rather than the global South, where many countries have too few resources to exhaust on training students of the past and rely on those from the North to provide such services.

The dominant language of archaeology globally is English: it is the most widely spoken and understood language in the world because of British success as a coloniser as well as the more recent political dominance of the USA as a majority Anglophone nation (another echo of politics as it emerged above). This also means that the majority of archaeologists from outside the English-speaking world are effective linguists, while those whose native tongue is English tend to be more deficient linguistically. It has led to debates—especially in non-Anglophone Europe—about the most effective manner of ensuring communication between archaeologists internationally. The leading international organisations—the IUPPS and WAC to which it gave an uncomfortable birth in 1986, the EAA (based in the Czech Republic), as well as other regional associations in other continents—have adopted English as the main language of communication, although they may also provide space for others: membership of these bodies is also high from Anglophone territories (see e.g. EAA 2022, 8). Kristiansen (2001, 43) pointed out in a discussion piece at a gathering in Poland the tendency for marginalisation of the archaeological traditions of non-Anglophone countries, especially smaller territories who do not share in a larger distribution of their language, to the detriment of global archaeology as a whole. The ensuing discussion (Kobyliński 2001, 44–46) saw only one departure from a consensus that English should be selected as a medium of wider communication, and that from one of the Anglophone attendees. To be an archaeologist on the global stage, then, is to be well acquainted with and have facility in English.

This discussion leaves us for the time being with the opening idea: that an archaeologist is anyone who does archaeology, whatever that may be.

Examining Archaeology as Practice

As can be seen from previous sections of this chapter, others have sought to investigate and comment upon aspects of archaeological practice, whether as a form of work (loosely defined and not limited to paid employment), or to establish or legislate as to what archaeology is and who can be called an archaeologist. Specific forms of archaeological practice have also been the subject of examination, in historical and other terms. Such work is represented by two distinct trends, although the topic of each is identical: one trend is that of guidance on how to conduct work in archaeology, often in the form of manuals; the other is that of commentary upon such work and how it is done, why that way, and what this means. The distinction is one I have used elsewhere (Carman 2002, 5–6) between writing about what *should* be and writing about what *is*. The difference is important but not necessarily evident: quite often statements of *should* masquerade themselves in terms of *is* in order to give more authority to what is said.

The distinction between a manual and a commentary can be most easily seen in comparing two books about fieldwork, both produced at the start of the third millennium CE. Roskams' (2001) Cambridge Manual in Archaeology on *Excavation* belies the limitations of its title: it covers much more than the process of digging with chapters also dedicated to pre-excavation survey, necessary preparations for on-site work, as well as methods of recording. As such it offers complete guidance for archaeological fieldwork including discussions of debates concerning appropriate styles of research. This is work very much in the mode of *should* because it provides detailed models of appropriate working practice. By contrast, Lucas (2001) specifically offers *Critical Approaches to Fieldwork* designed to "re-think the nature of the archaeological process" (Lucas 2001, 17) by asking questions about how excavation is done and how the experience of excavation is then reproduced in the form of written words and images. While Roskams' book is about how to represent accurately the encounter with material that is what fieldwork entails, Lucas is concerned with the role that the encounter with material plays in archaeological work itself. While Roskams' guide to how to do archaeology takes us from material to its representation, Lucas' theoretical commentary takes us in the reverse direction, back to the encounter with material which is at the core of the work. This discussion is not intended as a critique of Roskams' work or others of like kind (e.g. earlier texts on excavation, Barker 1982; on geophysical prospection, Clark 1980; or on field survey, Ordnance Survey 1973): on the contrary, such works are essential in delineating the standards to which archaeologists need to work in order to produce the outcomes required of our field and as guides to how to achieve them. The distinction between Roskams' book and Lucas' is simply in terms of intent: one provides answers; the other raises questions. Both however are concerned with exactly the same topic: what archaeologists do on site and the transformation of that experience into an output.

While this concern with outputs reflects Shanks and McGuire (1996) who, cited above, emphasise the role of the archaeologist as a creator of new knowledge, it also raises questions as to what is the *real* work of archaeology. New knowledge can be created in ways other than turning an experience of a material encounter into written texts and images, and that knowledge may not need to be disseminated for it nonetheless to be knowledge: knowledge that is possessed by one individual alone is still knowledge, after all. But this is not what is meant by any scholar when the talk of knowledge creation: for them it is always something to be passed on so that others may develop it and provide a deeper understanding among a wider community. It follows that creation of knowledge in forms that can be passed on to others is an essential component of the work of archaeologists: the question then becomes one of identifying the wider community to which that knowledge should be made available. This is a topic to be addressed further in this book, but it should be noted that this is also a topic among archaeologists themselves (and see below on the list of those to whom archaeologists owe ethical responsibility). As Connah asserts—in offering a book in no way intended as a manual on how to write about archaeology (Connah 2010, xi)—"Writing about archaeology is the archaeologist's lasting contribution to society" (Connah 2010, 1). The book proceeds to offer insights into the author's experiences of producing texts for others to read and the processes involved in doing so—including relations with editors and publishers and the different approaches to writing for different audiences. For those such as Connah, therefore, archaeology is not at all about the material encounter, or not fundamentally: it is *really* about creating new understandings of the world and making "the results of our investigations widely available, primarily to other archaeologists and scholars in related disciplines, but also to society in general" (Connah 2010, 9) and to do so by writing it down.

It is evident that the practices of archaeology are varied, and the skills needed by the archaeologist as varied. They involve the skills needed for fieldwork—not only to wield trowel, hoe or other implement, including as Lucas lists them "a machine to strip off the… soil, followed by hand digging with mattocks, spades, [and] shovels" (Lucas 2001, 1)—but also as set out by Roskams (2001) and others:

- financial and administrative planning in advance of work, and to allow it;
- preliminary examination of documentary sources, including maps and aerial photographs, as well as earlier reports of fieldwork on the site;
- fieldwalking and ground survey, plus prospection by remote sensing;
- use of photography, mapping and drawing to record findings, as well as record sheets to describe them;
- interpretation of the results of these various methods to guide both further action and understanding of what has been revealed;
- and an ability to turn all this into a written and drawn record that can be understood by others, to use for their own purposes.

Implicit in these skills is another—the ability to identify objects worthy of attention by archaeologists and those tell-tale signs in the earth that indicate the presence of human activity in the past. This is no mean skill, one that cannot be easily taught and comes mostly from experience: elsewhere I have called this 'having eyes to see with' (Carman 2006) and what a colleague describes as the archaeologist's 'superpower' (Henry Chapman, pers. comm.). Having said this, however, experience also reveals that quite often professional archaeologists lack this ability: others—such as amateur archaeologists or metal detectorists—are far more able to spot likely sites for investigation or identify objects. It is also the case that during fieldwalking, it often depends on the specialism of the walker as to what is identified on the surface: pottery experts see pottery, lithics experts see lithics, zooarchaeologists see animal bones, etc., and they tend not to see other kinds of materials outside their own purview.

As will also be seen further in the book, there are additional skills such as a capacity to work with others to achieve outcomes, and to form personal and professional relationships of rivalry, alliance and friendship that contribute to the undertaking and continuance of work. Overall, is an ability to judge what is appropriate and proper, and what is not, and indeed for others too. This takes us to the realm of ethical decisions, the topic of the next section.

The Ethics of Archaeology

As has been heavily hinted if not directly stated above, issues of ethics will run as an undercurrent and recurring theme in this book because they are, one way and another, foundational to the field. As an indication, Praetzellis—in conversation (quoted in Rathje et al. 2017, 142–143)—presents the archaeologist, and especially in the role of the cultural resource manager, specifically as an "ethical professional" and goes on to present himself as a "moralist" in his own writings aimed at a popular non-professional audience (Rathje et al. 2017, 154–155). It is indeed very difficult to keep any consideration of ethics out of any archaeological discussion as any attempt to define the field or oneself as an archaeologist turns on them (see also the US Register of Professional Archaeologists *Archaeological Ethics Database* [https://archaeologicalethics.org/]).

As a consequence and unsurprisingly there is a great deal of modern literature on ethics in the field, as well as Codes of Practice and statements of ethical behaviour to which archaeologists are bound by memberships of groups and associations, together with actual laws governing how particular types of material should be treated. I have treated with this material elsewhere (Carman 2002, 80–81; 2015, 74–75) but in summary, today it consists of owing particular obligations to various others, not all human, as shown in Table 1.1.

As indicated, these are quite recent developments—only since the 1970s. Archaeology has a longer history than this, and depending on the authors one reads, dates from as long ago as twentieth century BCE Egypt (Trigger 2006, 43) or sixth century BCE Babylon (Trigger 2006, 44) or the early to mid-nineteenth century CE (Daniel 1975). In whatever case, however, ethical concerns were until recently limited to non-existent: the retrieval of ancient material—in Egypt, Mesopotamia or Europe—was carried out by those who had a claim on it, and recent writers have emphasised the essentially colonialist nature of archaeology in the nineteenth and twentieth centuries of our own era (see Diaz-Andreu 2007) as Europeans in other parts of the world gathered material to feed European museums. As a consequence, or as a cause, there was no great concern before our own age for the welfare of Indigenous populations or for the preservation of remains to serve the interest of

Table 1.1 Professional responsibilities of archaeologists (from Carman 2015, 75)

	Associations/professional bodies in				
	USA			Europe	UK
Professional responsibility owed to:	SAA	AAA	ROPA	EAA	CIFA
The public	●	●	●	●	●
Colleagues		●	●	●	●
Archaeological record	●	●	●	●	●
The past					●
Students		●		●	
Clients		●	●	●	●
Laws		●	●	●	●

those inhabiting the areas from which they came. Instead, the purpose of archaeology was to confirm the natural and inevitable status of emerging nation states and the inevitable superiority of European civilisation as the heir to the great cultures of the past. There is an ethic at work here—one that serves the development of the concept of modernity (Thomas 2004) and its spread across the globe—but it is a self-serving ethic of the right of investigators of the past to treat the material they retrieve in whatever manner they see fit. It laid no responsibility upon the individual archaeologist except to qualities of scholarship and European concepts of rightness that do not extend to others; Thomas (2004, 235–238) consequently argues for a 'counter-modern archaeology' with a different ethic.

The majority of considerations of archaeological ethics have been written by and for archaeologists themselves (e.g. Green 1984; Vitelli 1996; Tarlow 2001; and as part of discussions recorded in Rathje et al. 2017) but occasionally others are involved. In particular, notice of archaeology has been taken by others interested in issues of ethics and how they are addressed in terms of archaeological practice in two interdisciplinary edited volumes (Scarre and Scarre 2006; Scarre and Coningham 2013) involving philosophers, lawyers and historians as well as anthropologists and archaeologists. Much of the discussion revolves around issues of claims to ownership and control over material—especially of cultural property in its various forms, especially as a form of appropriation by archaeologists (a specific focus of Scarre and Coningham 2013). However, issues of how archaeologists engage with those whose pasts they study is also a key concern, whether they be living communities or the dead, and what this means for how archaeologists conduct themselves and the status they claim or can achieve. This is emphasised in Trigger's (2006) magisterial *History of Archaeological Thought*: he too points out how recent is a concern with ethics which he links with a concern for the interpretation of others' pasts (Trigger 2006, 543–544) and disputes about the use of data to provide the basis for 'stories' about the past rather than a concern for 'hard' facts (Trigger 2006, 472). The general trend of ethical concern in archaeology has therefore shifted over its history from *internal* relations of archaeologists with other archaeologists to *external* relations with others with whom archaeologists engage.

The recently-published EAA Code of Practice (EAA 2022) for its members is one of the longer and more detailed statements of professional ethics, covering a wide range of aspects of archaeological work:

- The different sectors in which archaeologists work
- The treatment of human remains
- The evaluation of material and concern for its provenance
- Relations with Indigenous communities
- The repatriation and restitution of human remains and other cultural material
- Archaeologists as teachers and trainers
- Archaeologists as heritage professionals
- Safety and wellbeing in the conduct of archaeological work.

The EAA Code sits alongside a set of EAA Principles which also address some of the same areas and may cause some confusion in their interpretation, which is a problem for all such codes.

Elsewhere, ethics finds it way in to the discourse of the field. It emerges for example in approaches to education in the discipline: as a set of specific topics relating to the politics of claims on material (Kersel 2007); as an integral component of introducing students to theory (Leach 2007); and as difficult questions about archaeological practice (Allen 2007). Meskell (quoted in Rathje et al. 2017, 350), echoing Tarlow (2001), confirms her own objection to the codification of archaeological practice because it serves to confirm archaeological authority: she prefers that archaeologists should "sit back and listen a lot more" because she puts people above material (Meskell quoted in Rathje et al. 2017, 341). She is supported in this in the same volume by Hodder (Rathje et al. 2017, 126) who is concerned to maintain the rights of others involved (Hodder 2010). The alternative strand of discussion of ethics concerns the treatment of material rather than people. Wylie—introduced as "archaeology's philosopher of science" (Rathje et al. 2017, 93)—is expressly concerned with issues of stewardship and material that has been looted from sites (Rathje et al. 2017, 112–119), while Alcock confirms a specific interest in responsibility to research and the archaeological record (Rathje et al. 2017, 245). Neither of these (as Wylie confirms: Rathje et al. 2017, 118) prevents a concern also for those whose pasts are under study which explains the frequent complexity of ethical statements and the apparent contradictions they may—and often do—contain.

All this serves to indicate the many ethical issues with which archaeologists are now expected to engage as part of their daily activity. None of these issues is easy of resolution, reflecting the rise over past decades of professionalism in archaeology, a term (as indicated above) used in this book in two senses: as a description of the rise of the archaeologist as a paid specialist in particular activities; and as a standard of performance by all those engaged in archaeology, whether paid or otherwise. The specific ethical rules under which we work of course only apply to those of us who are members of particular organisations—hence the duplications in Codes (Table 1.1)—and only so long as we remain a member of that organisation, but they nonetheless stand as general guidance to all archaeologists everywhere of the kinds of responsibilities we owe to each other and the wider world.

Issues of ethics underscore so many of the aspects of archaeological work covered in this book. They apply to issues of fieldwork and the management of their objects of study, to how archaeologists engage with each other and the wider world, to how and what they think, and to what archaeologists produce and how. As stated above, ethical questions are not easy to answer, especially when different interests clash, which is why they are so central to any field of endeavour. It is something that makes archaeology interesting as well as important.

The Structure of This Book

This introductory chapter has hopefully set the tone for what follows, as each chapter examines a particular aspect of archaeological practice, building towards an overall idea of what it is to be an archaeologist and what it means to be one, and why it matters that archaeologists exist. Others of course have also made the argument for the significance of the field—notably two American authors, Barbara J. Little (2007) in her study of *Historical Archaeology—Why the Past Matters* and Jeremy A. Sabloff (2008) in his statement that *Archaeology Matters* in the form of *Action Archaeology in the Modern World.* For Little, archaeology matters because it adopts a questioning attitude to what we so often take for granted about ourselves, offers a form of public scholarship that emphasises the complexity and diversity of pasts (Little 2007, 158) and offers opportunities for civic renewal and restorative justice in a divided America (Little 2007, 159–163). Sabloff (2008) takes a different tack, reviewing some key themes—environmental sustainability, war, and spreading urbanisation—and how archaeologists can inform and lead action to create a better world. I disagree with neither author but I take a different approach in this book from both by looking at archaeology from the inside rather than directly at its relations with a wider world.

Chapter 2 focuses on what many consider to be the central practice of archaeology—that of fieldwork. It can be argued (and some do) that anyone who does not engage in fieldwork is not an archaeologist, whatever claims they make. It is true that all archaeologists spend at least some time in the field—one cannot understand the field without doing so—but that is by no means the end of things as further chapters will show. We are trained as archaeologists through fieldwork in the outside, including not only excavation but surface survey and the exploration of new places, but we also draw on documentary sources. And a lot more is done after the fieldwork itself is complete and elsewhere.

Archaeologists not only retrieve material and by doing so destroy the sites they work on (although see further in the book for challenges to this idea), but go on to create outputs. As Chap. 3 will indicate, they make pasts, presents and futures from the work they do by producing written texts and other forms pf presentation, including talks, reconstructions and museum displays. They engage with publics and create particular kinds of people—not least other archaeologists, but also others who may be friends or foes. And in doing so they produce ideas about how the world works, adopting or challenging the ideas of others. As a consequence, as Chap. 4 shows, archaeologists are heavily engaged in communications. They do this among themselves in various ways both formal and informal and to others by the same means and others. Communication means more than words—it is also in what archaeologists do and in what others say about what they do. The chapter will argue that if archaeology is about anything, it is mostly about communication (and see similar ideas introduced above).

But archaeologists do not work entirely for themselves, as Chap. 5 will demonstrate—they also work in the public realm in a number of capacities as public servants and as agents of the State and in other official organisations. As will be seen, this opens archaeologists to charges of exercising an authority that may not rightly belong to them; but from the perspective of archaeologists in these roles they are truly servants of a higher authority. It is a debate as yet unresolved and the position taken is one dependent on one's own situation.

As a public service, archaeology is inevitably governed by political forces and how politics is inevitably involved in the field of archaeology will be examined in Chap. 6. However, archaeologists are also able to strike back at forms of political control, to represent alternative forms of politics and (as Sabloff 2008 argues) to take political action themselves. There is a link here with other practices—with issues of employment of course, but also ethics as discussed above, with the products of archaeological work, with communication, and with issues of ideas and philosophy.

Ultimately, archaeology is a form of *social* activity and this is examined in Chap. 7. It will be seen how archaeologists relate to each other as archaeologists, as fellow professionals, and how they build and sustain communities. As a closed world of those with specific knowledges and expertise, like other such bounded communities, the archaeological community works in particular ways; but its members also interact with the wider world individually and collectively to maintain particular relations with the world.

It is this particularity of archaeology that makes archaeology what it is, and Chap. 8 will gather the various and interlocking threads from the previous chapters to identify what that may be. Ultimately the argument that emerges will be that archaeology is a set of practices that are worth retaining for the value they provide which goes beyond providing interesting entertainment, utility in addressing current problems or providing economic returns—not that these are unimportant in themselves—but offers more than the sum of these benefits. Archaeology, it will be shown, not merely *matters* (Little 2007; Sabloff 2008)—it is *important*.

References

Aitchison, K. 2010. Discovering the Archaeologists of Europe. *Archaologisches Nachrichtenblatt* 15: 103–110.

Allen, M. 2007. Scenarios for Archaeologists: A Teaching Tool. In *Archaeology to Delight and Instruct: Active Learning in the University Classroom*, ed. H. Burke and C. Smith, 245–251. Walnut Creek: Left Coast Press.

ARI. 2006. *Needs Assessment Conducted for the Register of Professional Archaeologists*. Rockville: ARI Inc.

Barker, P. 1982. *Techniques of Archaeological Excavation*. London: B.T. Batsford.

Binford, L.R. 1989. *Debating Archaeology*. San Diego: Academic.

Carman, J. 1996. *Valuing Ancient Things: Archaeology and Law*. Leicester: Leicester University Press.

———. 2002. *Archaeology & Heritage: An Introduction*. London/New York: Continuum.

———. 2006. *The 'Archaeologists' Eye' on Geography: Looking at landscapes of battle*. Paper presented at Council for British Archaeology Winter Meeting on Archaeology & Geography, London, 9th March 2006.

———. 2015. *Archaeological Resource Management: An International Perspective*. Cambridge: Cambridge University Press.

Chapman, J. 2000. *Fragmentation in Archaeology: People, Places and Broken Objects in the Prehistory of South Eastern Europe*. London: Routledge.

Childe, G. 1942. *What Happened in History*. Harmondsworth: Penguin Books.

Childe, V.G. 1951. *Man Makes Himself*. New York: New American Library.

Clark, A. 1980. *Seeing Beneath the Soil: Prospecting Methods in Archaeology*. London: B.T. Batsford.

Clarke, D.L. 1968. *Analytical Archaeology*. London: Methuen.

Cobb, H., T.J. Harris, C. Jones, and P. Richardson, eds. 2012. *Reconsidering Archaeological Fieldwork: Exploring On-Site Relationships Between Theory and Practice*. New York: Springer.

Connah, G. 2010. *Writing About Archaeology*. Cambridge: Cambridge University Press.

Cooper, M.A. 2008. This is Not a Monument: Rhetorical Destruction and Social Context of Cultural Resource Management. *Public Archaeology* 7 (1): 17–30.

Council of Europe. 1992. *European Convention on the Protection of the Archaeological Heritage (Revised) (ETS No. 143)*. Available at: https://rm.coe.int/168007bd25. Accessed 28 Nov 2022.

Daniel, G. 1975. *150 Years of Archaeology*. London: Duckworth.

Diaz-Andreu, M. 2007. *A World History of Nineteenth Century Archaeology: Nationalism, Colonialism and the Past*. Oxford: OUP.

Edgeworth, M. 1990. Archaeology as Practical Reason: The Perception of Objects in Archaeological Practice. *Archaeological Review from Cambridge* 9 (2): 243–251.

———. 2003. *Acts of Discovery: An Ethnography of Archaeological Practice*, BAR International Series 1131. Oxford: Archaeopress.

———., ed. 2006a. *Ethnographies of Archaeological Practice: Cultural Encounters, Material Transformations*, Worlds of Archaeology. Walnut Creek: Left Coast Press.

———. 2006b. Multiple Origins, Development, and Potential of Ethnographies of Archaeology. In *Ethnographies of Archaeological Practice: Cultural Encounters, Material Transformations*, Worlds of Archaeology, ed. M. Edgeworth, 1–19. Walnut Creek: Left Coast Press.

European Association of Archaeologists (EAA). 2022. *EAA Annual Meeting Business Meeting Reports 2022*. Prague: European Association of Archaeologists. Available at: e-a-a.org. Accessed 23 Nov 2022.

Everill, P. 2009. *The Invisible Diggers: A Study of British Commercial Archaeology*. Oxford: Oxbow Books.

Flatman, J. 2011. *Becoming an Archaeologist: A Guide to Professional Pathways*. Cambridge: Cambridge University Press.

Gathercole, P., and D. Lowenthal, eds. 1990. *The Politics of the Past*. London: Unwin Hyman.

Green, E.L., ed. 1984. *Ethics and Values in Archaeology*. New York: Free Press.

Hamilakis, Y. 2013. *Archaeology and the Senses: Human Experience, Memory, and Affect*. Cambridge: Cambridge University Press.

Hamilakis, Y., and P. Duke, eds. 2007. *Archaeology and Capitalism: From Ethics to Politics*. London: Routledge.

Hermon, S., and V. Vassallo. 2012. Preventive Archaeology: Towards a Technological Integrated Solution. In *Revive the Past: Proceeding of the 39th Conference on Computer Applications and Quantitative Methods in Archaeology, Beijing, 12–16 April 2011*, ed. M. Zhou, I. Romanowska, Z. Wu, P. Xu, and P. Verhagen, 177–185. Amsterdam: Pallas Publications.

Hodder, I. 1986. *Reading the Past: Critical Approaches to Interpretation in Archaeology*. Cambridge: Cambridge University Press.

———. 1999. *The Archaeological Process: An Introduction*. Oxford: Blackwell.

———. 2010. Cultural Heritage Rights: From Ownership and Descent to Justice and Well-Being. *Anthropological Quarterly* 83 (4): 861–882.

Holtorf, C. 2006. Studying Archaeological Fieldwork in the Field: Views from Monte Palazzo. In *Ethnographies of Archaeological Practice: Cultural Encounters, Material Transformations*, Worlds of Archaeology, ed. M. Edgeworth, 81–94. Walnut Creek: Left Coast Press.

Kersel, M. 2007. The Great Debate: Archaeology, Repatriation and Nationalism. In *Archaeology to Delight and Instruct: Active Learning in the University Classroom*, ed. H. Burke and C. Smith, 55–65. Walnut Creek: Left Coast Press.

Kobyliński, Z., ed. 2001. *Quo Vadis Archaeologia? Whither European archaeology in the 21st century? Proceedings of a European Science Foundation Exploratory Workshop, Madralin near Warsaw 12–13 October 2001*. Warsaw: Institute of Archaeology and Ethnology, Polish Academy of Sciences.

Kristiansen, K. 2001. Borders of Ignorance: Research Communities and Language. In *Quo Vadis Archaeologia? Whither European archaeology in the 21st century? Proceedings of a European Science Foundation Exploratory Workshop, Madralin near Warsaw 12–13 October 2001*, ed. Z. Kobyliński, 38–43. Warsaw: Institute of Archaeology and Ethnology, Polish Academy of Sciences.

Lazar, I., T. Kompare, H. van Londen, and T. Schenk. 2014. The Archaeologist of the Future is Likely to be a Woman: Age and Gender Patterns in European Archaeology. *Archaeologies: Journal of the World Archaeological Congress* 10: 257–280.

Leach, M. 2007. Perspectives from a Pot: Introducing Archaeological Theory Through Visual Interpretation. In *Archaeology to Delight and Instruct: Active Learning in the University Classroom*, ed. H. Burke and C. Smith, 168–179. Walnut Creek: Left Coast Press.

Levine, P. 1986. *The Amateur and the Professional: Antiquarians, Historians and Archaeologists in Victorian England 1838–1886*. Cambridge: Cambridge University Press.

Little, B.J. 2007. *Historical Archaeology: Why the Past Matters*. Walnut Creek: Left Coast Press.

Lucas, G. 2001. *Critical Approaches to Fieldwork: Contemporary and Historical Archaeological Practice*. London: Routledge.

Marshall, Y. 2002. What is Community Archaeology? *World Archaeology* 34 (2): 211–219.

McClanahan, A. 2006. Histories, identity and Ownership: An Ethnographic Case Study in Archaeological Heritage Management in the Orkney Islands. In *Ethnographies of Archaeological Practice: Cultural Encounters, Material Transformations*, Worlds of Archaeology, ed. M. Edgeworth, 126–136. Walnut Creek: Left Coast Press.

Naffé, B.O.M., R. Lanfranchi, and N. Schlanger, eds. 2008. *L'archaéologie preventive en Afrique: enjeux et perspectives*. Saint-Maur-des-Fossées: Éditions Sepia.

Newton, C. 1851. On the Study of Archaeology. Paper Read at the Oxford Discourse of the Archaeological Institute, 18 June 1850. *Archaeological Journal* 8: 1–26.

Ordnance Survey. 1973. *Field Archaeology in Great Britain*. London: Ordnance Survey.

Pearson, M., and J. Thomas. 1994. Theatre/Archaeology. *The Drama Review* 38 (4): 133–161.

Rathje, W.L., M. Shanks, and C. Whitmore, eds. 2017. *Archaeology in the Making: Conversations Through a Discipline*. London: Routledge.

Roskams, S. 2001. *Excavation*, Cambridge Manuals in Archaeology. Cambridge: Cambridge University Press.

Sabloff, J.A. 2008. *Archaeology Matters: Action Archaeology in the Modern World*. Walnut Creek: Left Coast Press.

Salerno, V., N. Mazzia, M. González, and C.P. de Micou. 2019. Archaeologists, Treasure Hunters and Collectors: Heritage in the Spotlight. *Heritage* 2 (1): 135–148. https://doi.org/10.3390/HERITAGE2010010. Accessed 6 Dec 2022.

Scarre, G., and R. Coningham, eds. 2013. *Appropriating the Past: Philosophical Perspectives on the Practice of Archaeology*. Cambridge: Cambridge University Press.

Scarre, C., and G. Scarre. 2006. *The Ethics of Archaeology: Philosophical Perspectives on Archaeological Practice*. Cambridge: Cambridge University Press.

Schnapp, A. 1996. *The Discovery of the Past: The Origins of Archaeology*. London: British Museum Press.

Schnapp, A., and K. Kristiansen. 1999. Discovering the Past. In *Companion Encyclopaedia of Archaeology*, ed. G. Barker, 3–47. London: Routledge.

Shanks, M., and R.H. McGuire. 1996. The Craft of Archaeology. *American Antiquity* 61 (1): 75–88.

Shanks, M., and C. Tilley. 1987. *Re-Constructing Archaeology: Theory and Practice*. Cambridge: Cambridge University Press.

Smith, L. 2004. *Archaeological Theory and the Politics of Cultural Heritage*. London: Routledge.

———. 2006. *Uses of Heritage*. London: Routledge.

Spaulding, A.C. 1960. The Dimensions of Archaeology. In *Essays in the Science of Culture: In Honor of Leslie A. White*, ed. G.E. Dole and R.I. Carneiro, 437–456. New York: Crowell.

Tarlow, S. 2001. Decoding Ethics. *Public Archaeology* 1: 245–251.

Thomas, J. 2004. *Archaeology and Modernity*. London: Routledge.

Tilley, C. 1989. Excavation as Theatre. *Antiquity* 63 (239): 275–280. https://doi.org/10.1017/S0003598X00075992.

Trigger, B.G. 2006. *A History of Archaeological Thought*. Cambridge: Cambridge University Press.

Vitelli, K.D., ed. 1996. *Archaeological Ethics*. Walnut Creek: Altamira press.

Wheeler, M. 1955. *Still Digging: Adventures in Archaeology*. London: Michael Joseph.

Woolley, L. 1937. *Digging Up the Past*. London: Pelican Books.

Chapter 2
Fieldwork

It is common for writers on archaeological fieldwork to begin with an anecdote from their own experience, either to introduce the specific aspect of such work they wish to examine or question, or to establish their own credentials as a legitimate commentator, or both. I am no different.

My first encounter on site emphasises the essentially physical nature of that encounter and its very material nature. My partner and I were volunteer diggers at a large hillfort site in southern England and after a few days marking finds—as a general introduction to such work and especially the team (everybody passed through the finds hut)—I was called upon to clear out a storage pit that had been half excavated, leaving a clear stratigraphic section. The pit was over a metre deep and although the site—being on chalk—was well drained, overnight rain meant that the linoleum protecting the overhanging pit edge was slippery. In hoisting myself clear of the edge my hand slipped and I fell forward to bounce against the section. The section—to the great relief of all watchers—was unharmed while I suffered a cracked rib and heavy bruising all down one side, which occasioned little sympathy. Luckily it did not prevent me from continuing on site where I hope I proved useful.

Another experience was less painful. This was a site in Hertfordshire, being excavated in advance of a housing development. As volunteers my partner and I were given an area of the site several metres wide as our own, which we trowelled in spits of two centimetres at a time. Our great success was uncovering the only clear dating evidence for the site—a coin of Roman Emperor Septimius Severus (193 to 211 CE)—but little else of particular interest. Until that is we came across what appeared to be a pattern of stakeholes. On investigating these they turned out to be very strange: they went straight down at first but then seemed to turn upwards again. The site supervisor confirmed the peculiar nature of these features and in the absence of any other ideas we continued as we had been, excavating in spits with the supervisor passing by periodically to see what other strange things we had found. As we deepened the trench the ground became spongy and we could feel it bouncing below us as we moved across it. Then one corner slipped into a void below it,

J. Carman, *Archaeological Practice as Politics and Ethics*,
https://doi.org/10.1007/978-3-031-69828-6_2

followed by other parts of our section. Rabbit burrows! The whole area of the site had been undercut by rabbits at some point in the past and we had so thinned the covering soil that it could no longer bear its own weight. Then we all went to the pub.

Of course these stories—whether you found then interesting or enlightening or not—do take the story of fieldwork to its latter phase—that of actual digging. As indicated in Chap. 1, there is more to the work of the archaeologist in the field than using a trowel or other implement to shift soil and make discoveries. This is preceded by other kinds of work and followed by others. This chapter will examine these and their role in the field of archaeology. They are not presented as models to be followed: this is not a 'how to' book (for that see e.g. Barker 1982; Roskams 2001; Drewitt 2011); instead the centrality of fieldwork to the field is what is emphasised here. Without it the other aspects of archaeology—to be looked at in the following chapters—would not carry the importance they do, or even exist.

The Significance of Fieldwork

Archaeology is a fieldwork discipline. That is to say, fieldwork is central to identifying archaeology as an activity in its own terms, as a field of endeavour distinctive from all others that are either not fieldwork disciplines or may be fieldwork disciplines but where that fieldwork takes a different form. These other fieldwork disciplines—geology, anthropology, sociology, and some branches of geography, for instance—all have their own particular focus in terms of what they study and how they go about it: geologists study rocks and landform, anthropologists other cultures, sociologists their own society, geographers how space is and has been used. Archaeologists study the residues of past human activity and this can only be found—as anthropologists and sociologists put it—'in the wild' rather than in archives or libraries. This is what distinguishes archaeologists from other students of the past: we retrieve and interpret what it is people in the past have bequeathed to us, deliberately or accidentally, in the form of objects and traces of activity. They are the products of fieldwork—we seek them out, which is what makes work out in the field specifically a form of *work* as discussed in Chap. 1. It is also what makes what we tend to refer to as *data* in reality *capta* because we actively seek it out rather than coming across it by happenstance (Chippindale 2000).

Whenever one chooses to locate the origins of archaeology—in twentieth century BCE Egypt, or fifth century BCE Babylonia, or nineteenth century Europe, or any other specific period—it is invariably an aspect of fieldwork that identifies the activity as archaeology rather than anything else: merely telling stories about a past is not enough, because there has to be some engagement with material to make it archaeological. The despatch of individuals by medieval and early modern Popes to seek out Roman remains (Schnapp 1996, 110 and 126) is regarded as a step towards archaeology precisely because it is a direct engagement with physical material remains and not just with their written products, even though the purpose was not purely scholarly and therefore not archaeological as we understand it today (Welch

2005, 292–294). Similarly, the survey and cataloguing work of Bure and Worm in seventeenth century CE Scandinavia (Carman 2015, 8–9) represents work short of modern archaeology and for a somewhat different purpose (Carman 2015, 10) but it is the fact that they had to work in the out of doors that allows them to be granted status as significant precursors of modern archaeology. The importance of fieldwork to archaeology as a distinct area of work is identified from the earliest examples of the discipline, as emphasised by Schnapp (1996) in his survey of antiquarian activity from the earliest examples to the emergence of what we today recognise as full-fledged archaeology. This form of fieldwork is however limited—it involves almost entirely only exploration, identification and inventory (Carman 2015, 6–11) rather than the forms of detailed site survey and investigation as we know them today: they fall into the broader and more general category of 'interest in the past' than a scientific concern for specifics. Excavation and a concern for stratigraphic and contextual relationships emerges later, although early practitioners can also be credited with a capacity to identify chronological relations and the importance of context, of which more below.

Apart from its status as a defining feature of archaeology and its precursors, fieldwork has other claims to significance for archaeologists. Fieldwork is the means *par excellence* by which archaeologists gather the data (or as stated above, *capta*) that is the source of the inferences they are then able to make about the past, and from which they are then able to construct interpretations of that data/capta and the past. As emphasised previously, it is a direct engagement with material from the past; but more than this it is an encounter that takes place in first instance *outside*, away from the confines of a building. Elsewhere (Carman 2006, 97) I have drawn a parallel between archaeologists engaged in fieldwork and building workers on-site or the crew of a ship at sea: all work outside and are also separated from the rest of the world. It is these three aspects—the encounter with 'stuff', doing so in the open air, and as a dedicated group detached from external influence—that gives archaeology its other specific qualities as a social activity (a theme we shall return to in Chap. 7). None of this denies the importance of the work on materials that takes place in museums or laboratories (see e.g. Richards and Britton 2020), but such analysis depends always on the retrieval of material in the field.

The significance of fieldwork to the discipline therefore lies along a number of simultaneous dimensions and is what makes it a defining characteristic of archaeology as a discipline in its own right:

- That it involves direct experience with material
- That it is the source of specifically archaeological data
- That the experience is one that takes place in the out of doors
- That it is a collective, social experience rather than an individual one (except quite rarely).

These are also, as readers will note, defining features of archaeology itself. Fieldwork is therefore fundamental to the field—and one reason why it is covered in the first substantive chapter of this book—because it encompasses all aspects of the field and from it all other aspects of archaeology derive. May (2012, 167) states it quite plainly: "Fieldwork is at the heart of the archaeological endeavour". As Cobb et al. (2012, 1–2) have discussed in the same volume, however, fieldwork as a set of practices assumes some basic distinctions between kinds of things: subject and object; nature and culture; the social and the material. Their task is to re-examine how the relationship between archaeological theory and archaeological practices play out in the context of on-site behaviours, and an interesting outcome of that re-examination is—as May (2012, 177) puts it — "archaeologists who work in the field are outsiders" because their on-site observations are so often entirely detached from the interpretation of what those observations *mean*; these interpretations are made by others who more usually are found in offices with access to the academic discourses to which the interpretations contribute directly (as observed by e.g. Holtorf 2006, 88; Wilmore 2006, 120).

It becomes apparent that there is a paradox at the root of archaeology as a discipline: that while archaeology is dependent upon fieldwork and relies upon its outputs to achieve its products, fieldwork is simultaneously subordinate to what else archaeologists do. All too often, it is also true that those who conduct the work in the field are also subordinate to those who turn their labours into academic statements about the past.

Fieldwork as Training

Fieldwork does more than produce materials for interpretation: it is the process that produces archaeologists and serves to distinguish seniority of rank and other marks of status between archaeologists (Carman 2006, 98–100; Bateman 2006, 76–77; Holtorf 2006, 88–89). As Everill (2007, 484) puts it: field training is frequently considered "more about 'becoming' an archaeologist (thinking, talking, acting and dressing the right way) than actually being taught to excavate properly". His diary of a university training excavation in the UK covers not only the sequence of events over the course of a day—what those involved were doing as part of an excavation team—but also the reactions to the experience of doing so from the different perspectives of those involved: students, supervisors, volunteers. His discussion exposes some crucial questions anyone leading such training must address: whether the purpose is that of training and education only, or as a contribution to a research project, or primarily as research with any training outcome as a useful but largely accidental by-product; and how to accommodate the differential abilities of supervisors, students and others to produce an outcome acceptable to all those involved. Perry (2004, 237) emphasises similar issues, such as:

> [the] ultimate goals [of the activity], the pedagogies employed and their effectiveness in preparing students for careers in archaeology. Significant issues to explore include the overall structure of field programs, student:faculty ratio and duration of instruction, the knowledge and aptitudes imparted to participants, how learning and skill acquisition is assessed, as well as the balance maintained between student instruction, research objectives and time and budgetary constraint

As training in particular skills, fieldwork covers especially techniques of survey and excavation, of recording and of artefact identification. The latter concern relates closely to the description by Wilmore (2006, 117) of creating a new category of small find on site—a 'narrative find'—because it contributed to discussions about the interpretation of the site without being an object of specifically archaeological interest, which emphasises the interactive nature of fieldwork, both of humans with materials and humans with other humans. It emphasises too the division of one aspect from another: rather than the training being seen as a holistic unity, survey work is separated from digging, recording from digging, and artefact and finds activity from survey and excavation. Wilmore (2006, 117–118) goes on to explain how "for many participants [a discursive distinction between excavation and survey] was assumed to exist and was used to explain why cooperation… was often difficult to achieve in practice".

As a form of experience, training very much also concerns other more personal aspects: the remembrance of participants' "first trowel, the way the dirt smelled, the endless hours of sifting and processing cultural materials and the social atmosphere" (Perry 2004, 236). As Wilmore (2006, 118) points out, participation in field training "is not simply the culmination of a series of past events but also a point in a participant's life history" and this is confirmed by Perry (2004, 246) who reports on participants' responses to a survey of fieldschool experience that it was very important or essential to them as an individual. As Everill (2007), Perry (2004) and Weisman and White (2011), among others, all indicate, experience in fieldwork training plays a large part in participants' decision whether to continue a career in archaeology. Perry (2004, 256) provides the reason: that the training excavation is likely to be the participant's first experience of involvement in a research project and is therefore formative in their understanding of what it is to be an archaeologist (echoing the thoughts of Everill 2007 recorded above).

Field training is of course intended to be primarily an *educational* process and this is an aspect well covered in recent commentary on the topic. Flatman (2015, 144) emphasises how little was published on the educational aspects of archaeological training until the 1990s. He is able to identify reasons why he believes such training may well have been poorer in the 1960s, 1970s and 1980s—especially the smallness of classes, the variability of skill among those responsible for the training, and the lack of any recognisable career path in archaeology. He goes on (Flatman 2015, 147) to advocate that archaeologists (the 'we' here) "formally contextualise our teaching, develop a formal archaeological pedagogy, and genuinely mix theory and method in terms of practical teaching". Flatman's ideas are also reflected in a report on the role of fieldwork in archaeological education (Croucher et al. 2008) which concluded that the fieldwork experience is valued, provides valuable transferable skills, but needs to be integrated with other learning on the wider programme of which it is a part. While other commentators specifically reference the inclusion of classroom engagement with theory and method as part of fieldwork training (e.g. Weisman and White 2011; Dufton et al. 2019; Pretty 2000; Hester 1963), Cobb and Croucher (2020, 25) go further to advocate research specifically in the pedagogical aspects of teaching archaeology "rooted in and emergent from the material realities of our profession, to ensure the sustainability of the discipline, to ensure we produce high-quality, socially responsible graduates, and simply because we owe it to our students and future generations of archaeologists". Such arguments support once more the idea that it is fieldwork—and especially as a first introduction to archaeology—that produces archaeolog*ists*, and this is the intention.

Driving this concern for pedagogy and the role and conduct of training is a recognition—reflected in Flatman (2015)—of the failure of archaeological education to produce professional archaeologists of requisite quality for the workplace. Top of the list of failings reported to Gibbs, Roe and Gojak (2005, 25) in Australia was the complaint that archaeology graduates suffered a "lack of practical/field experience"; second was their lack of skills in terms of commercial application, which is vital in a privatised system of archaeological provision. However, it is noticeable how much of the more recent literature around fieldwork training is concerned not with university students but with amateurs and volunteers in the sphere of public archaeologies (e.g. Grant 2014; van Vollenhoven et al. 2017) and with training professionals in such work (e.g. Guinness 2014). While much of such commentary concerns the introduction of participants to the practical aspects of survey, such as Grant's use of plane table survey (2014, 146–147) or van Vollenhoven et al.'s focus on site visit, excavation and recording by drawing (2017, 130–132), Derudas and Berggren (2021) examine also the use of 3D digital images to develop student skills in site recording, combining transferable skills in digital technology with the hands-on experience of digging the soil.

It is clear that training produces more than people with technical skills. It produces—depending on context and purpose—different kinds of archaeologists: those destined for careers in commercial or cultural resource management archaeology, others dedicated to community archaeology, and others yet who are members of non-vocational teams. It will also (although few will wish to emphasise this) distinguish those who are skilled in fieldwork and those who are not, and even those who may be best kept well away from any on-site activity for fear of the damage they may cause. But it is the overall experience of fieldwork that is its dominant aspect: that of becoming an archaeologist, and the encounter with the material that lies at its heart. Archaeologists do not however only work with material: as the next section will discuss.

Documentary Sources

The simplest distinction that can be made between historians and archaeologists is that the former study the past through written records and the latter through material remains. Like all simple distinctions, while it has some useful merit, it is not entirely true: historians are increasingly concerning themselves with aspects of the material (Finn 2018) while archaeologists have always had recourse to written sources as part of their own work. Despite the value of such material in identifying sites and their history, including the presence of structures now lost, and events that may leave an archaeological trace, guides to archaeological practice tend to say very little about them. Barker (1982, 219) provides a short statement of how registers of births, marriages and deaths may allow the identification of a site's history in terms of "ownership, tenancies, and the relationships and social standing of generation of farmers [which can then be] correlated with the buildings they lived in and worked from". Similarly, Roskams (2001, 50) provides a short section on such material (a mere 29 lines) and lists only written history, legal cases and place-names as potentially valuable. In contrast to both, the UK Ordnance Survey guide to fieldwork (1973) makes no reference at all to written sources.

Nonetheless, the use of written sources as an inspiration for developing on-site fieldwork has a distinguished pedigree in archaeology. A famous—or infamous depending on your point of view—example is Schliemann's (1976a [1881]; 1976b [1884]) reliance on Homer's *Iliad* and *Odyssey* together with later Classical writers such as Strabo and Pausanias as indicators of both the existence and location of the city of Troy (Sayce 1976, x-xi) which he then set about excavating together with time out to work at Mycenae and on Ithaka (Schliemann 1976a, 43–50). The use of documents is widespread in the field of historical archaeology—where it may also be called 'documentary archaeology' (Beaudry 1988) or 'text-aided archaeology' (Little 1992)—in a range of ways. Wylie (2006) and Little (2007, 29–31) argue for the development of a specifically archaeological approach to the use of written sources rather than a reliance on them to tell the history of a site. This is echoed by Small (1999) who argues against the 'tyranny of the text' in classical Mediterranean studies. Wylie (2006) proposes a range of uses of documentary sources for specifically archaeological purposes in the USA: using especially legal and property records, to find people and understand the historical circumstances in which they lived; and to locate material objects in the context of manufacture and use. This is similar to the uses for census records and probate inventories in the UK context (Tarlow and West 1999) while Little (2007, 29–31) cites examples of how archaeological investigation challenges established historical narratives to provide new understandings of events and contexts in the past. All of these commentators, and Small (1999), emphasise the differences between written documents and material record, not only in terms of form but also the purposes for which they have been produced and the very different contexts as well, giving rise to what may be complimentary data or contradictory data (Little 1992). Especially for Wylie (2006) but for others too, the use of written sources needs to be done carefully and with due regard to what it is that the archaeologist specifically wishes to gain from them: the question is always how they can contribute to the archaeological search for knowledge and thus privileges archaeology as a discipline over history.

Guides to field practice take more notice of other paper-based materials. The Ordnance Survey (1973, 3–5) emphasise the value of their detailed topographical maps to archaeologists for indicating the presence of features, as well as the contribution archaeological survey can make to their maps by securely locating surface features. They—together with others (e.g. Barker 1982, 33–34; Roskams 2001, 43–45) — also emphasise the value of aerial photographs (Ordnance Survey 1973, 13–15) for the identification of sites but Roskams (2001, 44) emphasises the danger of reliance on them since "sites which show up [on aerial photos] may be only those which have already been partially destroyed" and that the absence of indications at a location may indicate that there has been less disturbance meaning better preservation. As he puts it (Roskams 2001, 44): "where air photographs suggest dense occupation may be a guide as to where *not* to excavate, since all sites are heavily disturbed". Aerial photography is of course a misnomer here: although relevant for the early efforts from aircraft in the 1920s and 1930s and the current growing trend to use powered light hang gliders or drones carrying cameras, later developments including satellite imagery have taken such imaging beyond the air. The term 'photograph' is also a little misleading if it is assumed to mean coloured images on photographic plate or in digital form: imaging also includes use of infra-red or ultra-violent spectra (Dorrell 1994) and the rise of LiDAR (Historic England 2018) to reveal otherwise invisible ground features.

The former statements encroach on the topic of the next section since they relate to—and are usually described as forms of—survey, and this paragraph encroaches also on a later part of this book but is nonetheless relevant here. That is the use of the work of previous archaeologists to inform the development of an investigation strategy. For Barker (1982, 36) this is so the location of earlier trenches can be identified and recognised as recent interventions; for Roskams (2001, 50–51) it is a source of useful information about the site, akin to historical sources, but to be treated as carefully because incomplete or inaccurate or simply conducted very differently from one's own approach, so that results may not be meaningfully compared or correlated with one's own. Of course the use of such material is predicated on being able to locate it since it may date from some considerable time ago and may reside in a national archive—such as of a journal or society—or be hidden in issues of a defunct local journal or held in a private stock of local materials.

One further source deserves passing mention, although not strictly documentary—that of oral testimony. Wylie (2006, 22) values it as a source of local knowledge and memory, helping to place archaeological work in its wider local social context. For Roskams (2001, 51) it is more to do with memories of physical changes to the site, natural or otherwise, which may affect how the site is interpreted while under investigation. The difference here can be explained in terms of purpose and context: Wylie writes from the context of American historical archaeology which frequently engages with quite recent periods and living communities whose past is under study; Roskams from the European context of deeper pasts and a greater detachment from them. A further source is collaboration with communities whose past is under study (see papers collected in Matthews 2011) especially Indigenous communities who will have their own interpretations and narratives of their pasts (Horsthemke 2008).

The type of documentary source to be drawn upon, and the manner of that drawing, depends on the purpose—to prove its truth (Schliemann 1976a, 1976b), to challenge it (Little 2007; Small 1999), to inform on social context and past lives (Wylie 2006) or to inform an excavation strategy (Roskams 2001). Each is valid in its own right but its utility depends on the ultimate purpose of the project of which it forms a part. It may come at the outset of a project, to locate a site or a reason for investigation; or at the end, to provide material to confirm or challenge findings. But the simple distinction between historical and archaeological study is denied either way.

Survey

The term 'survey' as used here covers a number of quite different activities, not all of them necessarily survey in a narrow sense, some involving no more than the faculties humans are born with and others involving increasingly sophisticated tools and technologies. The latter—since it involves the search for specific features, usually busied and invisible to the eye—is also referred to as 'prospection', raising the spectre of the popular image of the archaeologist as a seeker after treasure (Holtorf 2004). The earliest examples of anything that may be considered archaeological survey are the activities of individual antiquarians rather than teams: names such as fifteenth century CE Cyrus of Ancona, sixteenth century CE Raphael, John Leland and William Camden, seventeenth century CE Johan Bure, Ole Worm, John Aubrey and William Stukeley, and others, tend to dominate in discussions as the early progenitors of archaeology. These individuals worked alone either on their own behalf or for a (frequently Royal) sponsor and recorded monuments *in situ* (very often among other natural features) and collected objects. All produced drawings and plans of monuments and so can be considered as the inventors of archaeological survey and its representation (see e.g. Ucko et al. 1991 for Aubrey and Stukeley at Avebury in England). Worm is also credited with the first systematic collection of archaeological objects, but his *magister opus* the *Museum Wormianum* was an eclectic mix of the natural, the made and the merely strange and exotic. The surveys of this pre-modern period served the purposes of collection rather than further investigation, the term 'collection' including here the recording of monuments rather than necessarily their removal to another location, although the purpose for some—such as Cyrus and Raphael — was indeed to identify material that could be used for other purposes (Carman 2015, 7). Among those in later centuries engaged in colonial activity, Markham (1871, 170–203) specifically describes the efforts of those officers of the British government from 1784 to 1870 who were concerned to identify and describe the monuments of ancient India. This is presented as part of his wider coverage of all kinds of survey work in the subcontinent from the seventeenth century CE onwards, noting (Markham 1871, 203) the unfinished nature of that work.

Markham (1871) includes topographical survey in his coverage, although limiting its interest to purely natural features and separating it from archaeology. By contrast, Barker (1982, 27) emphasises that the location of archaeological sites in their setting is vital, making the direct link between the work of the "geologist, the geographer, the pedologist, the climatologist or the ecologist" and condemning historians and geographers who take no account of the work of archaeologists (Barker 1982, 28). Contour mapping—such as by plane table (Grant 2014, 146–147) or such devices as a laser range finder (e.g. Hayawaka et al. 2007)—as well as putting the site in its landscape context will reveal the presence of features that may be worthy of further investigation by excavation and will give clear indications, if their form (and therefore period) can be identified, of the longer history of the site. It will also of course give a general idea of the area within which the site is located—high versus lower ground, slope and other factors affecting site location and access.

Fieldwalking (Barker 1982, 35–36; Roskams 2001, 46–48)—or pedestrian survey as it is also called—is a way of identifying sites by the presence of artefacts on the surface in the upcast from ploughing or other forms of earth disturbance. Put simply it involves individuals walking across the site, following set transects, and recording the presence of archaeological materials as they go. These can then be individually marked or collectively within a grid set out over the survey area. Banning et al. (2017, 471–473) outline the factors that may affect how useful such a technique may be—the experience and training of the personnel involved; the survey design and methods; the "obtrusiveness" of material, which are issues of colour and shape that make it stand out from the ground in which it sits (Banning et al. 2017, 472–473); and issues of visibility such as the colour of the ground, lighting, and impediments such as vegetation. In earlier work, Banning et al. (2006) tested the effects of such factors as the time spent in detecting artefacts, the distance over which detection took place, and the effect of falsely identified targets, and in further work (Banning et al. 2011) specifically examined the effect that the width of transects had on artefact identification. All of these showed the 'diminishing returns' attending insufficient attention to the difficulties of artefact identification, but other work also indicates how performance can be improved by combining fieldwalking with the results of global positioning (GPS) which locates each find in its precise location (Poirier et al. 2019) and electromagnetic induction (EMI) which allows correlation of artefact finds and densities with below-ground features (Trachet et al. 2017). A further factor in effective fieldwalking is an understanding of the way artefacts behave in the ploughsoil from which they will be retrieved because that indicates the relation of such finds to any underground features from which they derive (see e.g. Schofield 1991; Boismier 1997). Nonetheless, fieldwalking remains a valuable tool which can be conducted by those otherwise unqualified for more invasive techniques or where more technological methods—such as those discussed below—are not available.

Techniques such as EMI belong more properly to the realm of 'prospection', or remote sensing, which is the search for features by the use of devices that identify differences in the subsoil without actually penetrating the ground. Depending on the device used—magnetometry which measures differences in the soil from those created by the earth's magnetic field; resistivity which measures changes in electrical resistance; ground penetrating radar (GPR) which measures the delay between a pulse sent by the device and its return—and the conditions prevailing since some work better in some conditions, others in other conditions, it may be possible to identify sub-surface features worthy of further investigation. Such techniques have been used to good effect across the world since the 1950s (Roskams 2001, 52; e.g. Cozzolino et al. 2020) and in certain areas of archaeology are not only useful but essential since invasive techniques are not available due to issues of great sensitivity, such as the investigation of sites associated with the Holocaust (Sturdy Colls 2015, 143) or because physical access is impossible, such as for undersea regions (Gaffney and Fitch 2022). One alternative to fieldwalking that has been offered by Orengo and Garcia-Molossa (2019) is combining automated machine-learning with images from drones to identify scatters of potsherds. A further tool of increasing value to archaeology—especially in the field of Conflict Archaeology (Scott and McFeaters 2011) although also controversial, especially in territories where amateur or avocational activity on sites is prohibited or discouraged—is the metal detector which can be used to locate bullets from battles or to find material unidentified in excavation which is in the spoil heap. More esoteric, and possibly even more controversial (Graves and Poraj-Wilczynska 2009; van Leusen 1998), is dowsing—using rods of wood or metal held in the hands to locate underground features. Whereas magnetometry, resistivity and radar are the province of the professional expert in their use, the latter two techniques rely heavily on the expertise of amateur users to achieve results which is one reason for their controversial nature, as well as the fact that in the case of the metal detector it is frequently associated with illicit activity on sites (Thomas and Stone 2009).

It is clear that archaeologists engaged in fieldwork are not just diggers who retrieve material from the ground (as defined by e.g. Clarke 1968, 12–13 and referred to in the previous chapter)—they need to be versed in a range of approaches, all of which rely on a high level of skill and training, and some of which depend on complicated technologies. Beyond those mentioned here, Roskams (2001, 54–57) also mentions the chemical analysis of soils as an indicator of occupation activity. Overarching the use of such techniques, archaeologists also need to be able both to record the results appropriately and to interpret those produced by machine (Clark 1990, 132–157). Roskams sums up the value of such preliminary on-site work as the "variety of techniques [that] can be used to discover new sites, and to throw light on their internal configuration and physical character before formal excavation starts" (Roskams 2001, 60).

Excavation

The purpose of modern excavation is perceived to be the construction of site histories, based upon the stratigraphic relationships between the features that represent human activity—the digging of pits and postholes, the construction of walls and other structures, the laying of floors, the deposit of material, and clearance for new uses and new sets of activity. It was not always the case although an appreciation of how one thing must succeed another was evident from relatively early: as in the case of Stukeley's recognition of the diversion taken by the Roman road to avoid Silbury Hill, indicating that the monument must predate the Roman road. Here I shall discuss three interlinked aspects of excavation as a technique of archaeology: its specific purpose, which determines the style of digging; the structure of the archaeological record and how it is understood; and finally who and what actually does the actual digging. These vary from time to time and place to place, always bearing in mind that archaeology is a specifically Western approach to the study of the past which has been successfully exported to the rest of the world.

Excavation does not appear as a specific tool of archaeology until the eighteenth century CE, unless we include the efforts of rulers in fourteenth century BCE Egypt and fifth century BCE Mesopotamia to uncover and reconstruct older structures; but others have noted that their purposes in doing so were not archaeological as we mean the term today (Thomassen 2005, 219–220). As noted by Trigger (2006, 58–60), excavations from 1738 and 1748 respectively at Herculaneum and Pompeii, Italy, were initially for the purpose of recovering statues from the sites but developed into a more systematic approach that revealed the sites as cities composed of buildings, roads and other features. The initial purpose of such work—the retrieval of objects—characterises most excavations carried out from the eighteenth century well into the 19th which by modern standards could be considered no more than the butchery of sites. It was common across Europe in the eighteenth and nineteenth centuries for groups of individuals to dig into ancient features on their land—especially prehistoric burial mounds—as an afternoon pastime and display of physical prowess in order to discover objects. Only too frequently what would today be considered potentially important evidence would be cast aside as insignificant or simply ignored while its context was happily destroyed in the search for something of more immediate interest.

As Lucas (2001, 29) emphasises in the case of late nineteenth century archaeology—generally accorded the status of an era of significant developments in a modern 'scientific' archaeology in Europe and North America—that "sites… are treated as 'mines' for artefacts" rather than being excavated to reveal the stratigraphy that will allow an understanding of their construction. He shows by detailed discussion of their approach to excavation and understanding of what constitutes the archaeological record (Lucas 2001, 29–32) that major figures of the time, who are one way or another credited with an emphasis on stratigraphy, are primarily concerned with the retrieval of objects rather than an appreciation of context. Artefacts are diagnostic of particular groups of people at particular periods of history and can be used to establish chronologies, which as Lucas (2001, 28–29) argues was the main aim of the late nineteenth century and early twentieth century pioneers in the field, making them closer to antiquarians (Lucas 2001, 32) in their approach than to archaeologists of the twenty-first century. Nevertheless, the reporting of figures such as Waorsaae in Denmark, Petrie in Egypt or Pitt Rivers in England, as well as others in the Americas—measured section drawings and plans supplemented by reports on 'finds' which cover more than artefacts but also what would today be called contexts and features—was detailed and accurate, allowing others to build on their example and to develop a true concern for stratigraphic relationships. Lucas (2001, 37–38) ascribes the latter to Wheeler and his 'box' method of excavation which divided the area to be excavated into a grid of squares, each box of which would be dug to reveal an overall plan but leaving standing sections on the face of each box. Lucas (2001, 41) emphasises that for Wheeler and his contemporary Kathrine

Kenyon a stratigraphic layer was any individual feature, such as a pit: and he points out the difference in apprehension from modern British approaches which separate the fill of a feature from its interface with other features or the context in which it lies (Lucas 2001, 43).

For Lucas, the purpose of excavators from the late nineteenth century to the mid-twentieth century was to establish sequences of development rather than sites as overall archaeological entities (Lucas 2001, 54). He argues that the development in the 1960s of digging by 'phase'—based on an understanding of the site's development derived from a combination of plan and section—led in turn to the development of a new mode of representation of what was found by excavation, and thereby a new conception of the archaeological record, "no longer a physical site, made of made of dirt and sharp stones, no longer the location of buried treasure, but an abstract, immaterial structured set of lines, numbers and text" (Lucas 2001, 58). The latter is derived from the development of the so-called Harris Matrix (named for its author; Harris 1989) which develops a new form of recording of features, considering them as distinct features composed jointly of deposits of material (e.g. 'fills') and as interfaces between them and what lies adjacent (e.g. 'cuts'). This is a focus on individual actions: it does away with a focus on 'layers' as the main focus of identification and replaces it with single context recording, the ultimate purpose of the matrix being to show each feature in relation to others to which it can be firmly related and thereby construct the 'immaterial' model of the site used for subsequent analysis.

Lucas' summary of the development of excavation technique, while noting the continuities of tools such as typological and cultural classifications from earlier periods, emphasises the ultimate purpose of investigation:

> For Pitt Rivers and others of his time [from the 1880s], the archaeological site was a repository of objects which, if carefully excavated and linked to types of monuments, were instrumental in the construction of evolutionary typological sequences. For Wheeler, Kidder and their contemporaries [1920s to 1950s], the site became a repository of an artefactual assemblage indicative of a culture group, and, if stratified, the locus of critical information on the chronological changes within this assemblage. For us [today], the site is a repository of behavioural patterns, structured activities revealed through close analysis of contextual association within or between assemblages (Lucas 2001, 62).

Harris' work puts the emphasis of modern excavators on stratigraphic relationships as a means for constructing a site history detached from any other source. Archaeologists are credited with borrowing the idea of stratigraphy from geology—especially from Lyell's work published in 1830 (Harris 1989, 1)—although as both Lucas (2001, 29–32) and Harris (1989, 9) point out it may not really be until the early twentieth century that the idea becomes really significant. Harris sets out the 'laws' to which archaeological stratigraphy operates, three of which are drawn from geology and one of his own creation for the specific purposes of archaeology. The three from geology are:

- The law of superposition, which states that a layer placed above another is more recent than that other (Harris 1989, 30);

- The law of original horizontality, which states that any deposits will naturally tend towards a horizontal position; deposits which have tilted surfaces were either deposited that way or lie in conformity to a basin of deposition (Harris 1989, 31);
- The law of original continuity, which states that if a feature is seen in a vertical section then part of it must have been eroded or cut away (Harris 1989, 32);

The law developed exclusively for archaeology is that of:

- The law of stratigraphical succession, which states that a "unit of archaeological stratification takes its place in the stratigraphic sequence of a site from its position between the undermost (or earliest) of the units which lie above it and the uppermost (or latest) of all the units which lie below it and with which the unit has a physical contact, all other superpositional relationships being redundant" (Harris 1989, 34). This is in essence a development of the law of superposition which makes it directly applicable to archaeological sites and allows analysis in terms of constructing a site history.

Together these four principles explain what archaeologists should seek when excavating: the specific relationships between discrete entities in terms of which is earlier or later and which is contemporary with another. Whereas earlier excavators focused on the contents of the soil—deposits in Harris' terms—Harris distinguishes such deposits from the interface between deposits and other deposits or the context into which they were deposited. The interface—the surface or edge of a feature—is separate from the contents that lie within it, which is the deposit. Both require recording in the stratigraphic sequence. (This distinction is also similar although not identical to the distinction between 'monument' and 'event' in an archaeological archive: the former is any kind of object, site or structure—a 'thing'; the latter is any kind of human activity associated with it, whether of discovery or investigation).

The reduction of such issues to a set of laws does not of course make the identification and recording of archaeological phenomena or the actual understanding of the record any easier in practice. As Patrik (1985) argued, the theoretical debates of the 1980s and 1990s which went to the heart of the issue concerned two distinct views of the record: as a physical 'fossil' record or as a 'text' (Patrik 1985, 28); she sought reconciliation by treating one as what is found on site (akin to Harris' 1989 stratigraphic model) and the other as the post-excavation interpretive process, but that was to fundamentally miss the point of the debate since the issue was always one of what it is that archaeologists actually encounter and the two models are entirely antithetical. The textual model (explained and critiqued by Buchli 1995) has largely been abandoned in modern archaeological discourse but its alternative—Middle Range Theory—periodically re-emerges as a viable concept (e.g. Kelly 2011). The latter was intended as the necessary link between observations made in the field—e.g. of stratigraphic relationships—and interpretations in terms of culture. In practice it collapses to two concerns: ethnoarchaeological studies (e.g. Binford 1978); or issues of site formation (Schiffer 1976). Both of these relate directly to the way the archaeological is produced by processes both cultural and natural and invite a reverse reconstruction of the processes involved (called by Schiffer 'transforms' and inviting criticism from Binford as a 'Pompeii premise' of some kind of original pristine state: Binford 1981). These concerns are of course the reverse of Harris' (1989) approach but are by no means antithetical to it: indeed, they complement each other quite neatly.

It is noticeable that Harris' concern with stratigraphy, the Middle Range Theory/ text debate and indeed the 'Pompeii premise' argument all derive from the Anglophone regions. As Lucas (2001, 43) and Greatorex (2004) point out in different ways, the record is perceived differently elsewhere. Lucas (2001, 43) emphasises that a cut (for Harris, a vertical interface requiring note) would not be recorded as a stratigraphic event outside Britain, while Greatorex (2004) discusses the mutual incomprehension of excavation and recording methods between British and German archaeologists, and Barford (2004) and Carver (2004a) discuss the very different styles of excavation across the regions of a single country, Poland and Germany respectively. To complicate the issue, Carver (2004b)—drawing upon experience in Britain, North America and continental Europe—points to the 'fluidity' of soil, challenging the geologically-based model of stratigraphy derived from the study of rocks; in such a view 'intrusions' into deposits and 'disturbances' to stratigraphy are nothing of the kind, but "our assumptions are wrong" (Carver 2004b, 56). He adds that "we need to recognise that we are creating data, in that we are selecting which sets of data – which criteria – are relevant to our observations" (Carver 2004b, 58). In other words, what is observed depends on the expectations of the observer, a point also made by e.g. Hodder (1999) because archaeological work is (as also emphasised in the previous chapter) essentially an interpretive process and this applies as much in the field as elsewhere.

The previous discussion has immediate implications for an appreciation of the archaeological record. If we were to change the way we excavate we would inevitably also alter the form of the archaeological record because the latter is a product of our understanding. Indeed, the physical process of excavation itself has changed over time: the careful scraping of the soil most commonly engaged in today (and as pointed out at the beginning of this section) was preceded by eras of brute hacking at the soil to reach what it contained in the form of objects. And as Lucas (2001, 38) points out, even as foundational a figure as Petrie advocated on discovering a wall to excavate alongside it to see where it led, thus destroying any stratigraphical relationships that may have provided valuable data. This in turn leads us to the practicalities of actually shifting earth to discover what it reveals: and who and what actually does that shifting.

Lucas (2001, 1) conveniently describes what happens on many an archaeological excavation: "using a machine to strip off the [top] soil, followed by hand digging with mattocks, spades, shovels and trowels" to reveal the evidence that lay below. Not every country would use a machine to strip topsoil (I remember the shock among some American colleagues when first hearing of it!) and others abhor such heavy tools as mattocks; in others yet, a different tool from the trowel—perhaps a hoe—is used for revealing buried features. Nevertheless, the basic principle is the same wherever excavation is carried out. But as Lucas (2001, 3–5) explains, drawing upon others, and as stated above, excavation of any kind is quite a late development and even as late as the nineteenth century in Europe those considered as leaders in the developing field mostly relied on material collected by others. There was, however, a tradition of fieldwork in a wider sense: the topographic work of early practitioners mentioned in the section on Survey above, and the inventories of sites and materials created to support claims of ownership and control over that material. As Lucas (2001, 5) is clear to state, there was "a division between the fieldworkers and the intellectual elite, which more or less reproduces the same division between the person who collected the material and the museum or individual who commissioned the expedition". Indeed, he accuses authors on excavation practice (such as Alexander 1970) of not only grading "people according to their authority but includes objects such as trowels and excavators so that an impression is given of a hierarchy from human down to machine" with those in the lowest grade akin to tools (Lucas 2001, 55).

Everill (2012) charts the careers of 'invisible diggers' in Britain who were foundational to archaeology in the early nineteenth century, in particular the Parkers of Heytesbury who worked for Cunnington and Colt-Hoare. He also emphasises how the tradition of employing workers to do the actual digging continued well into the twentieth century. Despite being considered as 'pioneers' of excavation technique, those such as Pitt Rivers, Petrie, Woolley, or indeed Wheeler when overseas, were sparse in their wielding of the spade or trowel if indeed they ever touched one. Instead, the work of digging was left to skilled others whose names are rarely recorded and whose true contribution to the development of archaeological method and knowledge is therefore obscured. Taking an expressly Marxist approach, Mickel (2019) reveals the extent of debt owed to such workers in the Middle East: their ability "to recognize sites, stratigraphic transitions, and to lift fragile artifacts unharmed allowed for new discoveries and understandings that would have otherwise been impossible" (Mickel 2019, 197). She goes on to emphasise the degree of alienation such workers suffered, but also points out those means by which they asserted their own agency: "to do all of these things [as listed above] was not their only available option. They decided to dig, expertly, and the discipline as a whole benefited" (Mickel 2019, 199).

Above I referred to 'workers' but—as Lucas (2001, 7–8) emphasises in his discussion of gender in the field—for a large part of the history of archaeology, the term would be more properly 'work*men*' because of persistent gender divisions in archaeological work. As Gero (1994) has indicated, women are commonly relegated to aspects of archaeological 'housekeeping' in the laboratory or finds hut, rather than allowed to undertake the major projects addressing the 'big' archaeological questions. In similar mode, Moser (2007) has highlighted the essentially masculine culture of archaeological fieldwork. And as Lucas (2001, 8–9) also points out, the excavation is a realm of discipline—either by tight control and surveillance by the director and their supervisors, or by the mechanisms of "*pro forma* context sheets and other records which can proliferate excessively, and which are used to constantly check the quality and quantity of work" (Lucas 2001, 9). The hierarchical structures of earlier ages are thus continued into the present and the practical excavator—the 'digger' or the 'excavation technician'—placed lower in the hierarchy than the person directing the work. This theme is echoed by Everill (2012) who compares archaeological excavators with horticulturists (who also dig in the soil and are most likely to be educated to at least degree level), and by Berggren and Hodder (2003) in a review of commercial archaeology in the UK and USA. As the latter explain, despite shifts in method to allow more responsibility and independence to those who do the digging, such as single context recording which relies heavily upon the individual excavator's experience, knowledge and skill, "even in such cases the amount of authority that was delegated [is] limited, and there have been technological changes more recently, such as the use of digital recording, that may sometimes be used to again create a gap between excavator and interpreter" (Berggren and Hodder 2003, 425), thus linking the hierarchical structure with modes of discipline. As will be discussed further in a later chapter, professional archaeological organisations (e.g. CIfA in the UK and Ireland; ROPA in North America) distinguish between those of various grades of expertise with, for instance, 'practitioner' significantly lower than full member in the UK (https://www.archaeologists.net/join).

It is a paradox of archaeology that despite the centrality of fieldwork—especially excavation—to the discipline, those who conduct that activity are also those frequently given the lowest status, and at the worst treated as merely equivalent to appendages to the director of the work and to the tools (spades and trowels) by which it is conducted. Ultimately, although archaeology is *about* material culture and the archaeological record (hence Lucas' 2001 ultimate concern with issues of materiality), it is *done* by knowledgeable and skilled human beings, as the work of e.g. Everill (2012), and contributors to Edgeworth (2006) and Cobb et al. (2012), among others, emphasises. This theme of archaeology as an essentially *human* activity will be explored further in later chapters.

Exploration

The earlier section on survey ascribed the earliest activity to those who went—or were sent—to find what existed in the world around them. But what they were doing would just as easily fall into this section because there is a long tradition of archaeologists as explorers of other peoples' territories in exactly the same way as explorations of their own: for example, the Science News Letter (1924) had a section explicitly linking *Exploration, Archaeology and Anthropology* as essentially similar endeavours. It is also no accident that the names of leading organisations dedicated to archaeological investigation in the Middle East are the (British) Egypt Exploration Society (see e.g. Spencer 2007) and Palestine Exploration Fund (PEF; e.g. Hallote et al. 2012) together with the American Palestine Exploration Society. As Cobbing (2012, 180) outlines, the British PEF was driven by two competing aims: "popular religious, patriotic, and colonial enthusiasm.... [and] a strict employment of the latest scientific methodology and rationality in the construction and implementation of a hugely ambitious field research programme". As Cobbing puts it, the close connection that developed between the PEF and Cook's travel agency has "evolved into an enduring relationship between the archaeology, cultural and natural heritage of Jordan, and its modern tourist industry" (Cobbing 2012, 191). Similarly, Rekowska-Ruszkowska (2016) charts the connection between the search for archaeology to the exploration of Libya over the two hundred years to the early twentieth century. It is also evident in activities elsewhere from the middle nineteenth century. As Dyck (2009) records, interest in the archaeology of mid-western Canada began in the 1850s, although the East had been subject to archaeological exploration from the 1820s. Such work was originally conducted together with and in the same manner as investigation of geology, flora and fauna, and therefore tending to relegate the Native peoples of Canada as part of nature. Similarly, Rainbird (2022) outlines the role of explorers in discovering significant rock art sites in Western Australia from 1838.

The explicit connection of archaeological—or at least antiquarian — concerns with European expansion and exploration of others' territories is especially evident from Markham's (1881) review of the various surveys undertaken by the British in India from the seventeenth to nineteenth centuries. The majority of these are topographical one way or another—from coastal surveys from the seventeenth century onwards, to the surveys of viable trading and military routes, to detailed mapping and geological survey of the entire territory, and ultimately to the Archaeological Survey of India (Markham 1871, 170–203). The bulk of such work related to the recording and deciphering of inscriptions and the survey of upstanding buildings and other structures but it also led to the first efforts of the government to preserve ancient Indian structures (Markham 1881, 200–201). The ultimate outcome was the creation of the Archaeological Survey of India which remains responsible for archaeological work and preservation (Ota 2010, 89; Carman 2015, 22–23). The connection of archaeology with European explorers is further made explicit in the lives of such figures as Lady Hester Stanhope (Yegül 2022) and those who 'discovered' Nineveh in Mesopotamia in their travels either as independent adventurers (Stanhope) or imperial administrators (Layard, Canning, Rawlinson; Hoeks 2018).

Archaeological fieldwork is not the innocent pursuit we may like to make it. This chapter's overview is by means definitive or total: but the links of archaeology with European colonial expansion are evident from the close connection with exploration; and it is evident that fieldwork is highly structured in social terms, a realm of discipline and of class and gender division from the outset. Nonetheless, as citations in this chapter attest, archaeologists themselves are keenly aware of these aspects of their discipline and work to overcome them. Furthermore, they are aware that in terms of purpose the site of fieldwork is exploited for what it contains—for a large part of the history of archaeology, simply objects to be retrieved, and today more often as the repository of deposits to be investigated by their removal. Not that the simplistic claim that archaeology destroys its object is entirely true: as Lucas (2001, 40) points out, "in excavating a site or feature, we do not annihilate matter, it does not vaporise or disappear—it is merely displaced.... rather than destroying that which we wish to understand, we alter it". What we alter it to is the set of records that archaeologists create in order to understand what it is they have been working on. That is where the next chapter turns our attention: to what it is that archaeologists make by the work they do.

References

Alexander, M. 1970. *The Directing of Archaeological Excavations*. London: John Baker Ltd.

Banning, E.B., A.L. Hawkins, and S.T. Stewart. 2006. Detection Functions for Archaeological Survey. *American Antiquity* 71 (4): 723–742.

———. 2011. Sweep Widths and the Detection of Artifacts in Archaeological Survey. *Journal of Archaeological Science* 38: 3447–3458.

Banning, E.B., A.L. Hawkins, S.T. Stewart, P. Hitchings, and S. Edwards. 2017. Quality Assurance in Archaeological Survey. *Journal of Archaeological Method and Theory* 24: 466–488.

Barford, P. 2004. Excavation Methodology in Poland: A Cause for Concern? In *Digging in the Dirt: Excavation in a New Millennium*, British Archaeological Reports International Series 1256, ed. G. Carver, 117–127. Oxford: BAR.

Barker, P. 1982. *Techniques of Archaeological Excavation*. London: B.T. Batsford.

Bateman, J. 2006. Pictures, Ideas, and Things: The Production and Currency of Archaeological Images. In *Ethnographies of Archaeological Practice: Cultural Encounters, Material Transformations*, Worlds of Archaeology, ed. M. Edgeworth, 68–80. Walnut Creek: Left Coast Press.

Beaudry, M.C., ed. 1988. *Documentary Archaeology in the New World*. Cambridge: Cambridge University Press.

Berggren, Å., and I. Hodder. 2003. Social Practice, Method, and Some Problems of Field Archaeology. *American Antiquity* 68 (3): 421–434.

Binford, L.R. 1978. *Nunamiut ethnoarchaeology*. New York: Academic.

———. 1981. Behavioural Archaeology and the 'Pompeii premise'. *Journal of Anthropological Research* 37 (3): 195–208.

Boismier, W.A. 1997. *Modelling the Effects of Tillage Processes on Artefact Distribution in the Ploughzone: A Simulation Study of Tillage-Induced Pattern Formation*, BAR British Series 259. Oxford: Archaeopress.

Buchli, V. 1995. Interpreting Material Culture: The Trouble with Text. In *Interpreting Archaeology: Finding Meaning in the Past*, ed. I. Hodder, M. Shanks, A. Alexandri, V. Buchli, J. Carman, J. Last, and G. Lucas, 181–193. London: Routledge.

Carman, J. 2006. Digging the Dirt: Excavation as a Social Practice. In *Ethnographies of Archaeological Practice: Cultural Encounters, Material Transformations*, Worlds of Archaeology, ed. M. Edgeworth, 95–102. Walnut Creek: Left Coast Press.

———. 2015. *Archaeological Resource Management: An International Perspective*. Cambridge: Cambridge University Press.

Carver, G. 2004a. Method and Theory in German Archaeology: Preliminary Findings. In *Digging in the Dirt: Excavation in a New Millennium*, British Archaeological Reports International Series 1256, ed. G. Carver, 135–139. Oxford: BAR.

———. 2004b. Preliminary Reflections Upon the Unnatural History of Archaeological Stratigraphy. In *Digging in the Dirt: Excavation in a New Millennium*, British Archaeological Reports International Series 1256, ed. G. Carver, 35–61. Oxford: BAR.

Chippindale, C. 2000. Capta and Data: On the True Nature of Archaeological Information. *American Antiquity* 65 (4): 605–612. https://doi.org/10.2307/2694418.

Clark, A. 1990. *Seeing Beneath the Soil: Prospecting Methods in Archaeology*. London: Batsford.

Clarke, D.L. 1968. *Analytical Archaeology*. London: Methuen.

Cobb, H., and H. Croucher. 2020. Introduction: Valuing the Undervalued—Researching Teaching and Learning in Archaeology. In *Assembling Archaeology: Teaching, Practice, and Research*, ed. H. Cobb and H. Croucher, 1–25. Oxford: Oxford University Press.

Cobb, H., T.J. Harris, C. Jones, and P. Richardson. 2012. Reconsidering Fieldwork: An Introduction: Confronting Tensions in Fieldwork and Theory. In *Reconsidering Archaeological Fieldwork: Exploring On-site Relationships Between Theory and Practice*, ed. H. Cobb, T.J. Harris, C. Jones, and P. Richardson, 1–14. New York: Springer.

Cobbing, F. 2012. Thomas Cook and the Palestine Exploration Fund. *Public Archaeology* 11 (4): 179–194.

Cozzolino, M., Caliò, L. M., Gentile, V., Mauriello, P. and A. di Meo 2020. The Discovery of the Theater of Akragas (Valley of Temples, Agrigento, Italy): An Archaeological Confirmation of the Supposed Buried Structures from a Geophysical Survey. Geosciences 10, 161. doi:https://doi.org/10.3390/geosciences10050161l. Accessed 30 Jan 2023.

Croucher, K., H. Cobb, and A. Brennan. 2008. *Investigating the Role of Fieldwork in Teaching and Learning Archaeology*. Liverpool: Higher education Academy.

Derudas, P., and Å. Berggren. 2021. Expanding Field-Archaeology Education: The Integration of 3D Technology Into Archaeological Training. *Open Archaeology* 7: 556–573.

Dorrell, P. 1994. Ultra-violet and Infra-red Photography. In *Photography in Archaeology and Conservation*, Cambridge Manuals in Archaeology, 198–207. Cambridge: Cambridge University Press.

Drewitt, P.L. 2011. *Field Archaeology: An Introduction*. 2nd ed. London: Routledge.

Dufton, J.A., L.R. Gosner, A.R. Knodell, and C. Steidl. 2019. Archaeology Underfoot: On-Campus Approaches to Education, Outreach, and Historical Archaeology at Brown University. *Journal of Field Archaeology* 44 (5): 304–318.

Dyck, I. 2009. Retrospective: Canadian Prairies Archaeology, 1857–1886: Exploration and Self Development. *Canadian Journal of Archaeology / Journal Canadien d'Archéologie* 33 (1): 1–39.

Everill, P. 2007. A Day in the Life of a Training Excavation: Teaching Archaeological Fieldwork in the UK. *World Archaeology* 39 (4): 483–498.

———. 2012. *The Invisible Diggers: A Study of BRITISH COMMERCIAL ARCHAEOLOGY*. 2nd revised ebook edition. Available at https://www.academia.edu/4244219/The_Invisible_Diggers_A_study_of_British_commercial_archaeology_2nd_Edition_. Accessed 7th February 2023.

Finn, M. 2018. Material Turns in British history: 1. Loot. *Transactions of the Royal Historical Society* 28: 5–32. https://doi.org/10.1017/S0080440118000026.

Flatman, J. 2015. A Slight Degree of Tension': Training the Archaeologists of the Future. *The Historic Environment: Policy & Practice* 6 (2): 142–155.

Gaffney, V. L. and Fitch, S. (eds) 2022. Europe's Lost Frontiers: Vol. 1 Context and Methodology. Oxford: Archaeopress.

Gero, J. 1994. Gender Division of Labour in the Construction of Archaeological Knowledge in the United States. In *Social Construction of the Past: Representation as Power*, One World archaeology 24, ed. G.C. Bond and A. Gilliam, 144–153. London: Routledge.

Gibbs, M., D. Roe, and D. Gojak. 2005. Useless graduates?: Why Do We All Think That Something Has Gone Wrong with Australian Archaeological Training? *Australian Archaeology* 61 (1): 24–31.

Grant, K. 2014. Canals, Cake, and 6H Pencils: Starting Out in Community Archaeology. *Journal of Community Archaeology & Heritage* 1 (2): 137–154.

Graves, T., and L. Poraj-Wilczynska. 2009. "Spirit of Place" as Process: Archaeography, Dowsing and Perceptual Mapping at Belas Knap. *Time and Mind* 2 (2): 167–193.

Greatorex, R. 2004. Pan-European Archaeological Research: How Different Methodological Approaches to Research and Practice in Different Countries Might Affect the Interpretation of Evidence. In *Digging in the Dirt: Excavation in a New Millennium*, British Archaeological Reports International Series 1256, ed. G. Carver, 3–8. Oxford: BAR.

Guinness, A. 2014. Are They Being Served? Delivering Relevant Training for the Next Generation of Community Archaeologists and their Future Employers. *Journal of Community Archaeology & Heritage* 1 (2): 118–136.

Hallote, R., F. Cobbing, and J.B. Spurr, eds. 2012. *The Photographs of the American Palestine Exploration Society*. Boston: American Schools of Oriental Research with the Palestine Exploration Fund and the Harvard Fine Arts Library.

Harris, E.C. 1989. *Principles of Archaeological Stratigraphy*. 2nd ed. London: Academic.

Hayawaka, Y.S., T. Oguchi, J. Komoatsubara, K. Ito, K. Hori, and Y. Nishiaki. 2007. Rapid On-site Topographic Mapping with a Handheld Laser Range Finder for a Geoarchaeological Survey in Syria. *Geographical Research* 45 (1): 95–104.

Hester, J.T. 1963. A Training Program for Archaeology. *American Antiquity* 28 (3): 392–394.

Historic England. 2018. *Using Airborne Lidar in Archaeological Survey: The Light Fantastic*. Swindon: Historic England.

Hodder, I. 1999. *The Archaeological Process: An Introduction*. Oxford: Blackwell.

Hoeks, R., 2018. 'Many Great Treasures' of 'Great Beauty', or 'Crude and Cramped'? The Appraisal of 'Nineveh's Remains' by Austen Henry Layard, Stratford Canning, and Henry Rawlinson. *Bulletin of the History of Archaeology*, 28(1). https://doi.org/10.5334/bha-594. Accessed 9 Feb 2023.

Holtorf, C. 2004. Doing Archaeology in Popular Culture. In *The Interplay of Past and Present*, 42–49. Södertörns högskola.

———. 2006. Studying Archaeological Fieldwork in the Field: Views from Monte Polizzo. In *Ethnographies of Archaeological Practice: Cultural Encounters, Material Transformations*, Worlds of Archaeology, ed. M. Edgeworth, 81–94. Walnut Creek: Left Coast Press.

Horsthemke, K. 2008. The Idea of Indigenous Knowledge. *Archaeologies* 4 (1): 129–143.

Kelly, R.L. 2011. Why Did Binford's Middle-Range Program Outcompete Schiffer's Formation Process Program? *Journal of Archaeological Method and Theory* 18: 284–290.

Little, B.J., ed. 1992. *Text-aided Archaeology*. London: CRC Press.

———. 2007. *Historical Archaeology: Why the Past Matters*. Walnut Creek: Left Coast Press.

Lucas, G. 2001. *Critical Approaches to Fieldwork: Contemporary and Historical Archaeological Practice*. London: Routledge.

Markham, C.R. 1871. *A Memoir on the Indian Surveys*. London: W. H. Allen & Co.

Matthews, C.N., ed. 2011. Special Issue: Dynamics of Inclusion in Public Archaeology. *Archaeologies* 7 (3): 482–489.

May, S. 2012. Commentary. In *Reconsidering Archaeological Fieldwork: Exploring On-site Relationships Between Theory and Practice*, ed. H. Cobb, T.J. Harris, C. Jones, and P. Richardson, 167–180. New York: Springer.

Mickel, A. 2019. Essential Excavation Experts: Alienation and Agency in the History of Archaeological Labor. *Archaeologies* 15 (2): 181–205.

Moser, S. 2007. On Disciplinary Culture: Archaeology as Fieldwork and Its Gendered Associations. *Journal of Archaeological Method and Theory* 14: 235–263.

Ordnance Survey. 1973. *Field Archaeology in Great Britain*. Southampton: Ordnance Survey.

Orengo, H.A., and A. Garcia-Molossa. 2019. A Brave New World for Archaeological Survey: Automated Machine Learning Based Potsherd Detection Using High-Resolution Drone Imagery. *Journal of Archaeological Science* 112: 105–113.

Ota, S.B. 2010. Archaeological Heritage Resource Management in India. In *Cultural Heritage Management: A Global Perspective*, ed. P.M. Messenger and G.S. Smith, 82–98. Gainesville: University Press of Florida.

Patrik, L. 1985. Is There an Archaeological Record? *Advances in Archaeological Method and Theory* 8: 27–62.

Perry, J.E. 2004. Authentic Learning in Field Schools: Preparing Future Members of the Archaeological Community. *World Archaeology* 36 (2): 236–260.

Poirier, N., F. Hautefeuille, A.L.R. Ladeia, and É. Hautefeuille. 2019. Archaeotracker: la prospection pédestrie géolocalisé à haute resolution spatiale. *Les Nouvelles de l'Archéologie* 155: 5–11.

Pretty, K. 2000. Facts and Skills: Archaeology in Teacher Training. *Antiquity* 74 (283): 214–218.

Rainbird, M.P. 2022. Explorers and Researchers: Kimberley Rock Art Discoveries 1838–1938. In *Powerful Pictures: Rock Art Research Histories Around the World*, ed. J. Hampson, S. Challis, and J. Goldhahn, 126–135. Oxford: Archaeopress Archaeology.

Rekowska-Ruszkowska, M. 2016. *In Pursuit of Ancient Cyrenaica...: Two Hundred Years of Exploration Set Against the History of Archaeology in Europe (1706–1911)*. Translated by Anna Kijak. Oxford: Archaeopress Archaeology.

Richards, M., and K. Britton, eds. 2020. *Archaeological Science: An Introduction*. Cambridge: Cambridge University Press.

Roskams, S. 2001. *Excavation. Cambridge Manuals in Archaeology*. Cambridge: Cambridge University Press.

Sayce, A. H. 1976 [1884]. Preface. In Schliemann, H. 1976b [1884]. Troja: Results of the Latest Researches and Discoveries on the Site of Homer's Troy, 1882. Reprint edition. New York: Arno Press, v-xxx.

Schiffer, M.B. 1976. *Behavioural Archaeology: Principles and Practice*. New York: Academic.

Schliemann, H. 1976a [1881]. *Ilios: The City and Country of the Trojans. The Results of Researches and Discoveries on the Site of Troy and Throughout the Troad in the years 1871, 72, 73, 78, 79.*. Reprint edition. New York: Arno Press.

———. 1976b [1884]. *Troja: Results of the Latest Researches and Discoveries on the Site of Homer's Troy, 1882.*. Reprint edition. New York: Arno press.

Schnapp, A. 1996. *The Discovery of the Past: the Origins of Archaeology*. London: British Museum Press.

Schofield, A.J., ed. 1991. *Interpreting Artefact Scatters: Contributions to Ploughzone Archaeology*. Exeter: Short Run Press.

Science News Letter. 1924. 4(144), 5.

Scott, D.D., and A.F. McFeaters. 2011. The Archaeology of Historic Battlefields: A History and Theoretical Development in Conflict Archaeology. *Journal of Archaeological Research* 19 (1): 103–132.

Small, D.B. 1999. The Tyranny of the Text: Lost Social Strategies in Current Historical Period Archaeology in the Classical Mediterranean. In *Historical Archaeology: Back from the Edge*, One World Archaeology, ed. P.P.A. Funari, M. Hall, and S. Jones, vol. 31, 122–136. London: Routledge.

Spencer, P., ed. 2007. *The Egypt Exploration Society: The Early Years*, 16th occasional publication. London: Egypt Exploration Society.

Sturdy Colls, C. 2015. *Holocaust Archaeology: Approaches and Future Directions*. New York: Springer.

Tarlow, S., and S. West, eds. 1999. *The Familiar Past? Archaeologies of Later Historical Britain*. London: Routledge.

Thomas, S., and P.G. Stone, eds. 2009. *Metal Detecting and Archaeology*. Woodbridge: Boydell Press.

Thomassen, A.K. 2005. *Luxury and Legitimation: Royal Collecting in Ancient Mesopotamia*, Perspectives on Collecting. Ashgate: Aldershot.

Trachet, J., S. Delefortrie, M. van Meirvenne, B. Hillewaert, and W. de Clercq. 2017. Reassessing Surface Artefact Scatters. The Integration of Artefact-Accurate Fieldwalking with Geophysical Data at Medieval Harbour Sites Near Bruges (Belgium). *Archaeological Prospection* 24: 101–117.

Trigger, B.G. 2006. *A History of Archaeological Thought*. 2nd ed. Cambridge: Cambridge University Press.

Ucko, P.J., M. Hunter, A.J. Clark, and A. David. 1991. *Avebury Reconsidered: From the 1660s to the 1990s*. London: Unwin Hyman.

Van Leusen, M. 1998. Dowsing and Archaeology. *Archaeological Prospection* 5: 123–138.

Van Vollenhoven, A.C., K. Scott, and M. Harcombe. 2017. Learners' Encounter with Archaeological Fieldwork: A Public Participation Archaeology Account of the East Fort Archaeological Project. *Yesterday & Today* 17: 121–139.

Weisman, B.R., and N.M. White. 2011. A model Graduate Training Programme in Public Archaeology. *Antiquity* 74: 203–208.

Welch, E. 2005. *Shopping in the Renaissance: Consumer Cultures in Italy, 1400–1600*. London: Yale University Press.

Wilmore, M. 2006. Landscapes of Disciplinary Power: An Ethnography of Excavation and Survey at Leskernick. In Edgeworth, M. (ed.) *Ethnographies of Archaeological Practice: Cultural Encounters, Material Transformations*. Worlds of Archaeology. Walnut Creek: Left Coast Press, 114-125.

Wylie, L.A. 2006. Documentary Archaeology. In *The Cambridge Companion to Historical Archaeology*, ed. D. Hicks and M.C. Beaudry, 13–33. Cambridge: Cambridge University Press.

Yegül, F. 2022. Lady Hester Stanhope, a Monumental Arch, and Multiple Readings of a Triumph at Palmyra. *Journal of the Society of Architectural Historians* 81 (2): 175–196.

Chapter 3
Archaeology as Product

It is an axiom of archaeology that an excavation that goes unreported is wasted effort and indeed a direct breach of disciplinary ethics. This puts a focus on the responsibility of archaeologists to go beyond fieldwork to produce something that represents, even if it cannot replicate or replace, what has been destroyed or damaged—or as Lucas (2001, 40) puts it "altered"—by the efforts in the field. Archaeological work is therefore also aimed at the production of something that will persist beyond the finalisation of a particular project. But as this chapter will discuss, archaeologists produce more than reports on their fieldwork.

Among the physical products are indeed the archives and reports that derive from fieldwork, but depending on the kind of archaeologist one is, and the kind of archaeology that is practised by you, there are also journal articles and book chapters, books, television programmes, radio talks, blogs, vlogs and podcasts, internet websites, policy statements, consultancy documents, official returns, financial statements, and a host of other written, printed and broadcast materials (see e.g. Connah 2010). A crucial product of archaeology—and one too often overlooked—is the way in which archaeology especially produces categories of person, as I have indicated elsewhere (Carman 2015, 178–180)—as archaeologist or not, as ally, threat or as an irrelevance (as further discussed in Chap. 7). Beyond all these are the ideas and understandings and new knowledges that archaeologists produce as a result of their work, including (as touched on in Chap. 2) insights into their own work and the structure of their field of endeavour (see specifically in relation to the products of archaeological work e.g. Harlan 2010; Roth 2010). And overarching all this archaeology has the capacity to explore new ways of understanding and appreciating the world and what it is to be a material human being among all other the material things in the world.

All these different products will be addressed in more specific detail in what follows, but the essential point here is that none of this would be possible if archaeology were not primarily a *social* endeavour rather than a mere exercise of technical skill. As indicated in Chap. 1 and reiterated in Chap. 2, archaeology—above all

J. Carman, *Archaeological Practice as Politics and Ethics*,
https://doi.org/10.1007/978-3-031-69828-6_3

else—is a community that spans the globe: and it is notable how easily one archaeologist specialising in a particular branch of the field can discuss their work meaningfully with any other archaeologist specialising in a different branch and from an entirely different part of the world. This theme will be further addressed in Chap. 7 but it is also an idea that runs throughout the book.

Archaeologists as Makers

Chapter 2 emphasised the place of the archaeologist as a retriever of things from the earth—either in the form of physical objects or as forms of information or data/capta. This puts the archaeologist in the position of a parasite on others' residues rather than a more positive contributor to the world. But archaeologists do more than take—they also contribute to the world in a number of ways. As is indicated below, archaeologists make pasts, presents and futures by their work; they produce writing, drawings and other forms of representation; they engage with various kinds of others; and they produce understandings of the nature of what it is to be human in a material world.

It was noted in Chap. 1 that archaeology has been described as a craft activity (Shanks and McGuire 1996), a point developed by Shanks in other work:

> [craft combines] the technical, ethical and aesthetic in a unified practice of embodied knowledge. Archaeology as craft is argued as involving dialogue with the material past, and with client community, incorporating pleasure and learning, having interests in authority and the responsibility of the archaeologist-craftworker. Creative and poetic, as well as ethical, it is described as a sensuous receptivity to the past (Shanks 1992, 168).

In such a conception, archaeology is inevitably a creative process—akin to wood-, metal- or textile-working—involving a particular class of non-verbal practical knowledge (Edgeworth 2006, 38) involving the use of particular tools which are manipulated in skilful ways to alter material. This act of creation has consequences—as Harris puts it (2018, 37), "the material worlds we excavate create room for certain forms of humanity [and pasts, presents and futures] and not others"—a point that will reappear below. But that does not deny the act of creation that lies at the heart of the field.

Archaeologists also create things other than archaeology: they can make art. The 'incavation' project (Holtorf 2004) initially reversed the usual process of archaeology by burying material rather than excavating it, but also carefully recording the process and mounting that record as part of an exhibition (as well of course as then publishing a paper on it). As Holtorf (2004, 3–4) notes incavation is not "about faking archaeological evidence, about making archaeology appear a futile exercise… Instead, what is incavated is archaeological evidence in itself, and incavations are thus about the very possibility of archaeology". In similar vein, and as noted in Chap. 1, Pearson and Thomas (1994) examined the similarities between theatre performances and the archaeological process, especially as a challenge to the then-emergent notion of archaeology 'as [static and fixed] text', offering us the chance to "see performance as a contemporary experimental practice in understanding processes of cultural transformation, as an 'experimental archaeology of events'" (Pearson and Thomas 1994, 154). While the latter is not expressly art in itself, both these instances introduce us to the idea of archaeology itself as an artistic endeavour rather than a more narrowly 'scientific' one (although that by no means diminishes the creativity involved in any branch of science). Archaeological monuments have also provided artists with opportunities to display their work in an interaction with those monuments (e.g. Schülke 2000) to provoke responses around topics such as "alienation, provocation, myth, past and feelings" (Schülke 2000, 271)—ideas also present among Shanks' (1992) concerns.

Archaeologists produce things but, as indicated above, they also produce categories of person. As Chap. 1 showed, the concept of 'the archaeologist' is a slippery one and depends upon specific contexts for its application: in Anglophone usage, anyone who does anything archaeological can be included, and that is the option chosen for this book because it is the broadest application; but elsewhere the title of 'archaeologist' can only be applied properly to those with particular formal qualifications and accreditation. In claiming the title of archaeologist, one is also claiming a special authority to interpret pasts, so any others who also claim a relationship to pasts but lack that formal recognition can be, and often are, dismissed as less worthy, less reliable and certainly less qualified. This is especially the case for those who also retrieve material from the earth: pot-hunters, treasure seekers, metal detectorists and amateur collectors are all deemed by archaeologists as 'dangerous' because they challenge the special authority we grant ourselves and by their acts of retrieval may prevent us from investigating and interpreting the material they collect. Chapter 2 indicated the divisions within archaeology between 'diggers'—too often regarded as mere technicians—and those who interpret what the excavators retrieve and record (see e.g. Everill 2012). But we also grant authority to others whom we deem worthy: especially descendants—cultural or biological or both — of those who made the objects we retrieve and who may be able to provide insights into the cultural context from which they derive (e.g. Smith and Wobst 2005); and to officials who, in return for recognition by us as legitimate, grant to us the status we require to execute our work (see e.g. Thurley 2013). As people with a recognised position we are therefore able to create the other categories of person with whom we can choose to engage, ignore or challenge. We thus create the world we inhabit by establishing the limits within which it exists for us.

Productions of Pasts, Presents and Futures

That archaeologists produce understandings of the past—or of pasts—is a general assumption that many will agree with, and for many of those who do this will be understood to be all that archaeologists do. But of course we work in the present and for many who see archaeology as a political process (an aspect to be developed in

Chap. 6) the view is forward to the future as well as backward to the past. There are two main strands to this aspect of archaeological work: investigations of material cultures of the past, present and future; and an interest in temporalities of various kinds. The two aspects can interact and are frequently mutually implicative, and are therefore by no means mutually exclusive, so that work in one leads to the other and untangling them can be tricky. Nonetheless, they represent a key product of archaeology.

An explicit concern for producing and representing pasts is sometimes offered as a form of 'time travel' (Petersson and Holtorf 2017) whether in the form of the use of digital technology to create interactive experience, as an educational tool, as re-enactment and re-creation, via representation on screen, or simply by the use of imagination (and see a version of this for children: Frid and Holtorf 2021). Petersson (2017) discusses this in terms of the benefits of 'anachronism'; as he puts it:

> The conscious use of anachronism makes explicit that regardless of how we interpret the past – or the future – it is formed by the present. This fact… is extremely evident around us. We continuously see new perspectives on [the past] emerging, perspectives that change the past in favour of our own views, thoughts and wishes…. [such as] evidence of female power, queer perspectives, diversity, sustainability and the role of children and healthcare in the past (Petersson 2017, 285).

Holtorf (2016) considers similar themes in terms of what he calls the quality of 'pastness'—that is the ability of a place or object to provide an experience of the past in the present, so that "environments in the real world may successfully employ fictitious elements that saturate them with pastness and provide very historical experiences indeed" (Holtorf 2016, 34). He argues that there are implications for the understanding of temporalities of recreated pasts:

1. *Perceived temporal experience*: [which] is not dependent on actual age, but on the perception of pastness among particular human audiences.
2. *Context of history*: [which] is necessarily anachronistic and needs to be understood in terms of the functions, benefits, and values it provides for contemporary societies.
3. *Everybody is competent* to assess whether or not a particular item is successful in evoking the past in a meaningful way (Holtorf 2017, 126; italics in original)

Clearly such ideas link to how the past is represented to audiences (a theme addressed below and in Chap. 4) but the point here is that these offer tools for the analysis of direct and tangible links between different times, especially *then* and *now*.

In considering the distinction between a *then* and a *now* it is worth digressing for a moment to consider how archaeologists need to understand time. For any student of the past, regardless of how time actually works as part of the physical time-space continuum, time has to have three qualities or no sensible study of the past is feasible. Time must:

- Be fluid; always moving and never static, otherwise we would be stuck in a permanent present with no past or future and there would be no change, which is the measure by which we judge the passage of time;
- Be unidirectional; always moving forward from past through present to future, for any backward movement or looping makes identification of pasts, presents and futures impossible;
- Move at a constant rate; for if the 1000 years between 1 CE and 1000 CE were only as long as the 100 years from 1900 CE to 2000 CE (or *vice* versa) then any assessment of date or rates of change becomes impossible.

I emphasise that whatever relativity theory, string theory, quantum theory or any other model of the cosmos may indicate, for the archaeologist time *must* have these qualities to make archaeology possible. Without them, there is no distinction between past, present and future, and no possibility of establishing any system of dating—either relative or absolute.

A significant example of how archaeology creates pasts is the so-called 'Three-Age System', devised originally for Denmark by Christian Jürgensen Thomsen in the early nineteenth century and subsequently extended to the rest of Europe, the Middle East, with variants in the Americas, Africa and Asia. This was an important element in the development of a 'scientific' archaeology based especially on developmental theories of evolution, and in the development of cultural-historical approaches to the deep past. Today, we are familiar with the sequence of cultural and technological development that places lithic technology prior to metallurgy, and, within the latter, in Europe at least, copper and bronze technologies prior to that of iron. Later scholars developed the idea further: dividing the age of stone into a palaeolithic and a neolithic, and later still inserting a mesolithic to carry us over the transition between the two. Subsequently, the ages of copper, bronze, and iron all gained their regional subdivisions, too. Thomsen's story is well known and has been told many times (e.g. in Daniel 1975; Schnapp 1996; and especially Trigger 2006, 123–129). In 1816 Thomsen was appointed to catalogue the collection made by the Danish Royal Commission for the Preservation and Collection of Antiquities. He did so by starting with what he called 'closed finds'—deposits of objects that could reasonably be inferred had been made at the same time. By identifying which objects repeatedly appeared with which in such sealed contexts, he was able to infer which objects were in use (or at least available to be buried) together. The further study of changes in the style and form of objects in relation to their context of discovery allowed him to devise a chronological scheme for all of Danish prehistory. In all there were five stages rather than three: an early stone age, a later stone age, a bronze age, and an earlier and later iron age (the last of which continued into known historical times). First published as a guidebook to Scandinavian antiquity and translated almost immediately into German and then into English in 1848 (Ellesmere 1848), Thomsen's work was spread by his compatriot Jens Worsaae who influenced, among others, Daniel Wilson in Scotland (the inventor of the term 'prehistory') and Swiss investigators of the so-called 'Lake Dwellings' in that country. It went on to influence the rise of prehistoric archaeology across the globe. The scheme and its methodology fitted neatly into ideas about human social evolution and led the way to the construction of culture histories, an approach that would dominate much European and Anglophone archaeology until well into the twentieth century.

The long-term and ongoing significance of Thomsen's scheme is that for the first time there was a methodology for the construction of a relative chronology on the basis of material remains alone. Previous schema had relied upon biblical or classical references for their linkages, but his was one that had no need of external reference. Instead, it relied upon making connections between assemblages on the basis of their coappearance in the same context. The study of stylistic change—from relatively simple to more complex designs and their variations—provided a 'fine tuner' to allow the idea of gradual change over a flow of time rather than sudden alterations in cultural form. In other words, it was an entirely 'material culture' approach to a material culture issue: none such had existed previously. In many ways it represents the first example of 'archaeology as archaeology' (as advocated, for example, by Clarke 1968; Hodder 1986) rather than as an offshoot of historical, classical, biblical, art-historical, or folklore studies.

The connection with the creation of temporalities hinges on ideas of the experience of the past in the present, which in turn depends on being able to distinguish them. This is what contributors to Petersson and Holtorf (2017) were ultimately exploring, and Shanks (1992) has focused upon in his own work, distinguishing the formal academic approach of the archaeologist with the encounter with places evocative of the past. As he puts it: "the past is indeed stirring and evocative, but such feelings are separate from the study of the material past within an academic discipline" (Shanks 1992, 7). He advocates instead recognising archaeology as a craft activity (as seen in Chap. 1) which is always one of creation, an idea subsequently developed by Holtorf (2004)—albeit independently—in the idea of the 'incavation' whereby items are placed into the ground in an act of artistic creation rather than excavated from it as is more conventional in archaeology (as we have seen above). This is an effective reversal of the relationship of past to present: as Simonetti (2018) argues, archaeologists see the past as *below* and *underneath* them, and the further down the more distant in the past, but incavation places the present and future (or at least objects representing them) in this underneath. The subsequent act of retrieval reverses the reversal of chronological relationships to restore an archaeological normality; but it serves also to emphasise that the process of archaeology is to create temporalities, not to be subject to them.

Harrison (2016, 170) is clear that "an archaeological engagement with the present and recent past has involved an acknowledgment that archaeology constitutes a material and discursive intervention in the present". His reference here is to developing work since especially the 1980s—but some was earlier—which draws upon the archaeologists' skill in engaging with material culture (as discussed and emphasised elsewhere in this book) to examine our own present (see also Harrison and Schofield 2010). Examples include Graves-Brown's (2007) study of traffic islands, Harrison's (2009) of the UK Welfare State, and Schofield's involvement in studying microplastics in the oceans (van Sebille et al. 2019) and personal protective equipment in the context of the Covid crisis (Schofield et al. 2021). Similarly, in examining the consequences of the Fukushima disaster of 2011—involving an earthquake, tsunami and subsequent nuclear plant meltdown—Schlanger et al. (2016, 421) emphasise that:

> archaeologists can show that catastrophes are rarely, if ever, terminal, and that the decompositions they entail, whether above ground or below, widespread or *in situ*, are often recurrent and multiscalar [and that therefore] the material record as a whole is a dialectical combination of absence and loss.... [so that] archaeology can also confirm that time matters.

In these various ways, archaeologists of our own contemporary context create the possibilities of futures as yet unknown by revealing the complex relations of pasts to presents as reflected in and indeed created by interactions with the material.

Harrison (2016) particularly emphasises the role of archaeology as an aspect of heritage work in developing his argument about future directions, and others concur (e.g. Borck 2018). For Harrison, heritage work is inevitably a set of "Future-Assembling Practices" (Harrison 2016, 171–172) because heritage *management* practice is about identifying and mitigating risk and threat—identifiable risks and threats to particular objects of concern—and then addressing them in a manner that at least attempts to create a future state for that object. This means that the future condition of the object is envisaged as a part of that work—and it opens to scrutiny questions about the "temporalities… produced by different forms of heritage, and how… these articulate with the production of particular futures" (Harrison 2016, 172). For Borck (2018), writing from a specifically radical libertarian position, archaeology is a 'chronopolitical discipline' whereby "archaeological sites are being mobilized not just to legitimize the [present], but to create a future history where alternative power structures—egalitarian, non-state, Indigenous, pre-colonial—seem impossible to achieve; or worse, are forgotten" (Borck 2018, 229). Högberg et al. (2017) take a similar view: the future, they argue, is a constant referent in heritage work but this work is undertaken in a "constant, rolling present" (Högberg et al. 2017, 644) rather than an engagement with what futures might look like: accordingly, while the work creates futures, it does so as a replica of the *now*. Harrison's (2016) work, Borck's (2018) and that of others (e.g. Enqvist 2014; Nilsson Stutz 2018) all seek to overcome the creation of mere presents in the future by presenting alternatives for futures in the present. If anything confirms the capacity of archaeologists to create temporalities, it is this ability to think, write about and present versions of archaeology in terms of pasts, presents and futures.

Writing and Representation

Connah (2010, 1) asserts that "Writing about archaeology is the archaeologist's most lasting contribution to society". He goes on to quote Sinclair (1989, 161) that "archaeologists… write; [and thereby] we create our subject". Both—among others (e.g. Hodder 1989; Spector 1991; Chippindale 1996; Joyce et al. 2002; Joyce

2006)—are thereby keen to emphasise the significance of writing to the discipline and its centrality in the archaeological enterprise. Such writing is diverse in form: it is not only the types of writing (report, article, newspaper piece, blog, etc.) that are various but the styles as well. There are the solidly written but unreadable journal papers and technical reports—for specialists only—but also the books for the student or educated wider public, which are not identical but somewhat more accessible, and then also for the professional archaeological readership sometimes actual *stories* such as narratives of an object retrieved from excavation (Spector 1991) or Praetzellis' detective novels set on site—actually thinly disguised archaeological textbooks (Praetzellis 2003, 2011).

The very different types of written product from archaeology not only serve different audiences but are also designed to present a different version of archaeology. Connah limits his concerns to only three main types of output: the monograph; the journal; and miscellaneous 'others' which include "general interest popular magazines,… newspapers, [and] guidebooks" (Connah 2010, 46) among others not specified. Each of these broad categories is further subdivided. The monograph divides into the research report, the scholarly synthesis, the textbook, the collection of edited chapters around a single theme, sometimes in the form of a *Festschrift* celebrating the influence of a revered individual, the proceedings of a conference or conference session, and finally the book for the general reader. Journals are divided into the period or regional or technical specialist periodicals or those intended for a wider non-specialist readership: the contents of the former may well be subject to a process of scientific peer review which confirms the highly specialist nature of the work. To the list of 'others' already specified he adds "travel literature, encyclopaedias, exhibition catalogues" and writing for visual outlets such as television and the internet (Connah 2010, 57). He is ambiguous as to the status of the reports of archaeological consultants and commercial fieldwork units, which as he points out, have a "typically limited circulation" (Connah 2010, 46), are conventionally described as 'grey literature' (Seymour 2010) and therefore perhaps technically not 'published' but are nonetheless important forms of archaeological written product.

The scientific specialist publication—whether monograph, journal article or consultancy or commercial site report—is designed to present a vision of archaeology as a solidly academic, even scientific, pursuit: 'scientific' here sometimes used in the same sense as a 'hard' science but more commonly in the sense of an organised system of knowledge, created and shared by those with expertise and an understanding of a specialist realm. The student textbook has two main aims: to inculcate the newcomer into the specialist knowledge that is required to practise in the field, and to present archaeology as an interesting and exciting field of endeavour that can serve also as a career. The forms of general readership outputs—whether monograph, popular magazine article or guidebook—are designed to invite wider non-professional interest in archaeology and therefore ultimately to provide public support for the work of archaeologists (Cleere 1984; McGimsey 1984; and see below). Each of these purposes is different and although each serves to the benefit of the field and those who work in it, they nonetheless each require a different approach, content and style of writing.

Issues of style matter. Hodder (1989) compared the style of so many modern fieldwork reports with those of an earlier age. While the former are often so dry as to be almost abstract, conveying information as if it involved no human action or any choices made by those working on site, the latter are shown to be more concerned with the personal involvement of the author: indeed the chosen examples are more like personal letters than impersonal technical descriptions. Chippindale (1996) goes further to analyse the changing styles involved in producing a report on his own work on rock art in Australia. In arguing that "knowledge resides also in *how* it is expressed, and then also in *how* any one reader receives that account and integrates it with a larger frame of experience.... [so that] means of expression *transform* what is expressed" (Chippindale 1996, 48), he identifies four "moods" that deliver different kinds of information to the reader:

- "the common body of knowledge shared by specialists in the field;
- the particular attitudes and interests that are not shared;
- the new information or analysis reported, which should not be much affected by who chanced to execute the work;
- the several choices these researchers made in their study, in the kinds of information they chose to gather and the analyses they chose to pursue" (Chippindale 1996, 51–52)

and each of which requires a different mode of expression and use of words. The delivery of common knowledge requires much use of the verb 'to be'—something *is*, others *are*, and so on—and as such are statements to be taken as generally accepted truths about external reality (Chippindale 1996, 49). The delivery of information about disputed issues requires different language: someone *concludes*, another *argues*, yet another *contends;* all imply doubt and uncertainty about the general acceptability of the points made by others (Chippindale 1996, 50). The new observations made by the researchers producing the report are presented as neutrally and impersonally as possible, and as a collective enterprise, so not as totally objective but not as mere subjective impressions: there is agreement among the observers of what they saw, giving it some measure of authority (Chippindale 1996, 50–51). The choices made in collecting the observations and recording what was encountered are delivered in a manner similar to that of disputed material: the emphasis is on collective decisions and the language in the active voice, so that we are told they *chose, emphasised, believe*, placing the emphasis on the researchers' own presence and direct engagement in the work (Chippindale 1996, 51).

Chippindale's (1996) analytical focus is upon a published report originally presented at a conference rather than a 'grey' report (or as Börjesson 2015 prefers, 'professional literature'). Börjesson's own focus is particularly upon such professional reports derived from development-led archaeological work in Sweden, which she characterises as a 'boundary object' providing "a bridging role between archaeologists and other stakeholders of archaeological information [which serves to] not only organize field data but also influence relations between archaeologists, cultural heritage management professionals, land owners, and developers" (Börjesson 2015, 1162). As others (e.g. papers in Seymour 2010) point out, the perceived division between academic archaeologists and field practitioners is reflected in the attitude to such 'grey' documentation which is "often considered to be inferior to other kinds of publications" (Roth 2010, 338) simply because it is the product of archaeology conducted in a particular economic and social context. As Roth (2010) explains, such reports are seen as failing in three respects: that of quality control; that of ease of access; and that of scholarly interaction between practitioners and academics. The first of these reflects a common lack of recognition by academics of innovation in technique represented by reports, and a frequent specific lack of reference to theory, although such claims are overstated. The second reflects a widespread lack of suitable repositories and limited circulation, so the reports can be difficult to track down, issues which can nonetheless be addressed by establishing suitable central and regional repositories as in some countries. The third reflects the tendency for different communities of archaeologists to gather in different contexts, although this is increasingly mitigated by conferences and other scholarly events that deliberately attract both academic and professional attendees.

Börjesson's (2015, 1171–1172) analysis also highlights the frequent lack of citation of academic literature in professional reports, often because access to that literature from their own workplaces is limited, which is one factor for doubts about their quality. Harlan (2010, 284) presents a different view: while some "reports are replete with references to the academic [literature] the reverse seldom holds". As Chppindale (1996, 49) explains, referencing others' work "is itself an aspect to structuring knowledge; as well as serving the reader with guidance about further sources, it gives cause for the authority the writer claims [because they constitute] reports based on known evidence, published and available". It follows that to be taken seriously, and following Roth (2010), professional reports and other 'grey' literature need to be both referencing academic literature *and* be referenced by that literature in return. A further factor affecting both access and scholarly acceptance is that "knowledge production in development-led archaeology still seems largely confined to the national borders" (Börjesson 2015, 1174) so cross-border reference is difficult if not impossible, and reports themselves may be limited in their potential impact because comparison or complementarity with other work cannot be made.

The location of distinct archaeologies within the borders of separate states is a product of the well-established historical connection between archaeology and nationalist ideology. Indeed—as mentioned in Chap. 1—the emergence of the institutions of archaeology as a discipline and the nation state as the dominant form of political organisation are inextricably linked (Kohl and Fawcett 1995; Diaz-Andreu 2007). Any population wishing to establish their right to self-determination as a nation state will call upon their archaeological past to do so as in Ethiopia (Zena 2018) and Siberia (Mikhailov 2013). Similarly, populations within established nation states who wish to declare their separateness will do so—as in Brittany (France) (Dietler 1994, 593–595) or Catalonia (Spain) (Alonso 2013)—and so will states making claims on another's territory, as following late nineteenth century conflicts between states in South America (Gänger 2009). In such cases, and despite its international and indeed global scope, archaeology is presented as something different *here* from *there*. But it can also be used to unite otherwise disparate groups, albeit not at the global level. As Dietler (1994, 595–596) and Gramsch (2000, 11) note, a number of major exhibitions in the 1980s and 1990s sought to present a vision of a unified Europe in the past as a precursor to—and justification for—the establishment of the European Community (EC) and (its later manifestation) European Union (EU), both essentially political entities. As Tzanidaki (2000, 26) points out, the EU showed no concern for cultural matters before 1981 (25 years after the creation of the EC) and that concern has taken different forms over the period since, mostly concerned with culture as an economic force. The wider discussions to which Gramsch (2000) and Tzanidaki (2000) contributed were about the idea of a common European-ness that could be located in the archaeological record: and both they and others remain sceptical both of the idea and the role of archaeologists in its construction. Willems (2000), commenting on the two papers, emphasises the overarching value of debates among archaeologists at the European—and indeed world—level, to overcome the problems caused by difference while maintaining diversity. In such debates, archaeology presents itself as diverse and, as elsewhere, self-critical and, crucially, also of international (rather than mere national) significance.

Of course archaeologists do not only present themselves in words but also in images. Conventionally, these take the form of on-site activity: in an active role such as excavating a feature, using survey equipment, or explaining a site to visitors. They may also be more passive: as in displaying an object to an unseen observer. Such images are generally treated as self-explanatory and uncontroversial because they are of the mere day-to-day practices of the discipline but they can be very revealing. There is for instance the image placed on the cover of Cooper et al. (1995) of General Augustus Henry Lane Fox Pitt-Rivers directing excavations at South Lodge Camp: here everyone is entirely still, the General having dismounted from his carriage—seen in the background—while his workers hold their positions in or around the excavated ditch, tools in their hands at the ready. Here the class relations at work in the organisation of archaeological fieldwork—as discussed in Chap. 2—are graphically shown. Compare this with the image on the cover of Barker (1982) of two young men recording an excavated section with careful measurement and completion of paperwork. Here both appear equal in status and are crucially active rather than still. To some extent the images differ simply because of changes in photographic technology (one is monochrome, the other in colour; one requires stillness because of exposure length, the other does not) but underlying them is a significant social change: in the later image excavators are presented as skilled individuals in their own right, rather than subservient labourers; and while in the older image the impression is simply one of manual labour to move soil, the later is one of preparation to carefully record revealed features.

Shepherd (2003) discusses images of African workers from apartheid South Africa to reveal the discriminatory attitudes they display. He criticises the general run of on-site photographs as follows:

> what we find in site reports are stylised shots of individual artefacts (strategically lit and arranged against neutral backgrounds), and carefully composed shots of archaeological deposits (brushed, tidied, squared-away and labelled). These form a class of imagery from which coworkers and assistants are edited out, along with extraneous items of equipment, signs of camp life, collapsed sections and misplaced artefacts, in fact, any signs of production or failure (Shepherd 2003, 350).

The images on which he focuses are instead those of people—some anonymous, some identifiable—which reveal these generally hidden aspects of the working environment, such as the camp where they live and (especially since the notional excavator is White) those who do the actual work on site. In similar vein, Morgan (2014) has analysed the use of film to record archaeological work. She divides them into a number of categories:

- expository, "the more traditional, stand-alone, didactic expository style of documentary, using 'voice-of-god narration' and expert interviews" (Morgan 2014, 329)
- direct testimonial, or 'video diary', "a summary of current finds and conditions on site" (Morgan 2014, 331)
- phenomenological, "concerned with granting the viewer the gaze of an archaeologist.... to record the sensory components of archaeological fieldwork" as experienced (Morgan 2014, 333); and
- impressionistic, investigating "the interdisciplinary space between art and archaeology" (Morgan 2014, 332).

Each of these serves a different purpose in representing the archaeologist, in the same way other forms of representation—including texts—do. The first is the archaeologist as a professional and expert seeker of knowledge and archaeology as a realm of that specialist expertise. The second and third are the archaeologist as direct recorder of experience and archaeology as an interesting and exciting locale of that experience. The fourth goes beyond the third to present the archaeologist as artist—the creator not only (or even especially) of knowledge but of something that has not been encountered before: not a parasite who relies on the material of the past to supply their material, but a transformer of their particular experience into something entirely novel for consumption by others (and see also Schülke 2000). It is this self-image in particular that justifies and supports the claim of this chapter of archaeology as a realm of production.

Archaeologists and the Public

The production of texts assumes a readership for those texts, and while much archaeological output (especially technical and professional reports) is aimed at other archaeologists, other products are also aimed at a wider non-specialist audience (a point further developed in Chap. 4). While definitions of what is meant by 'the public' in the discipline vary—and in some cases are never provided—there are perhaps two that offer a good startpoint: Merriman (2004, 1–2) who distinguishes between archaeology as a public service, most usually conducted for the State (further discussed in Chap. 5), and archaeologists' engagements with communities of individuals; and McManamon (1991) who provides a convenient list of those categories of person comprising the public. Merriman (2004, 5–8) provides two models of engagement with communities of individuals: a *deficit* model whereby it is assumed that non-specialists are deficient in knowledge and understanding and require suitable education (akin to the arguments of Cleere 1984 and McGimsey 1984); and a *multiple perspective* model which offers the professional archaeologists' as just one way of understanding the world among a range of others. McManamon's (1991) list derives exclusively from the American experience but can be translated into more global terms as:

1. A general public at large
2. Those working and studying in education
3. Legislators and politicians
4. Professional administrators; and
5. Indigenous peoples

While McManamon's numbers 2–5 are specialist audiences requiring a particular manner of approach, number 1 (the general public) is a residual category that presumably excludes those other defined groups. Nonetheless, it emphasises the potential diversity of audiences, each of whom would also require different kinds of approaches, publications and outputs to suit their needs and wants.

Developing McManamon's (1991) approach, albeit frequently without acknowledging that they do so and not always explicitly, others have sought to define the kinds of publics that they engage with. Merriman (2004, 12–15) chooses to designate them as 'stakeholders', defined in terms of the range of type of individual "who has rights to own and interpret the material remains of the past" (Merriman 2004, 12). By contrast, Marshall (2002, 216) defines two kinds of community with which archaeologists engage:

> The first consists of people who live locally, either on or close to a site. Such communities are defined in the present and are largely about people's relationships to their place of residence.... The second kind of community consists of descendants and includes those who can or choose to trace descent from the people who once lived at or near the site.

The first type may have no claim of descent from those who made the archaeology, while the second may no longer live locally; but the two types of community may also be contained in the same group of people, and Indigenous populations are a special case of descendant communities. Other communities of interest include amateur collectors, metal detectorists and looters (e.g. Renfrew 2000; Bland 2004; Qin 2004; Thomas 2012; Brodie 2012; Kersel 2012), educationalists (Frost 2004; Kehoe 2012), the media (Ascherson 2004; McDavid 2004), exponents of 'fringe' and 'alternative' pasts (Schadla-Hall 2004), and of course archaeologists themselves (Darvill 2012; Kristiansen 2012).

As well as defining the types of public with which archaeologists engage, those involved in Public Archaeology are also concerned to identify the kind of work involved. For Moshenska (2017) these involve: working *with* the public on projects; work conducted *by* the public in the form of local amateur groups or independent scholarship; work *in* the public sector (Chap. 5); formal education; making archaeological work accessible *to* public audiences; offering a popular vision through various media; and 'academic' public archaeology which is the study of Public Archaeology and how it is done. Skeates et al. (2012) offer a perspective especially from that of Moshenska's 'academic public archaeology', examining the history of public engagements of various kinds, how academics study engagements with publics, how archaeology as a public service is done, and different kinds of engagement with different audiences. The latter comprises a range of approaches from the simple delivery of information, through closer engagement and consultation with communities on archaeological matters, to collaboration with communities, to community control and so-called 'democratic' archaeology (as described by Smith and Waterton 2009, 15–16). Each of these is different in form and ultimate purpose, requires distinctive skills, and yet they are often confused especially by those who (driven by a deficit model of publics) are concerned with public education and outreach rather than full collaboration.

Nonetheless, most concern in the field of Public Archaeology is with how best to work with or for the types of community identified by Marshall (2002), and especially with groups generally considered 'excluded' from archaeological concerns (see e.g. Schücker 2012). In particular this relates to Indigenous groups in the Americas (e.g. Rubertone 2008; Gnecco and Ayala 2011), Australasia (Smith and Wobst 2005), far North Europe (Harlin 2019), and elsewhere (e.g. Pratap 2009; Eze-Uzomaka 2000; Varouchakis 2017) but also concerns issues of class in North America and Europe (e.g. Mrozowski 2006) and subaltern racial and ethnic groups (e.g. Cuddy and Leone 2008). All of these can be considered as 'descendant' communities rather than 'local' (*per* Colwell-Chanthaphonh and Ferguson 2008) even when also locally resident but it is their status as a descendant community that mostly takes precedence in such concerns. However, as recent work emphasises, disadvantage frequently applies to people not because of one aspect of their identity but to a number at a time: issues of gender, ethnicity, class and other factors intersect requiring a recognition of "polythetic, overlapping/intersecting social group identities, experiences and power dynamics" (Spencer-Wood and Trunzo 2022, 50). For Spencer-Wood and Trunzo (2022, 23) intersectionality theory leads to a capacity to 'unpack' such generalising concepts as 'gender', 'race', 'sexuality', 'class' or 'race' to include also such factors as "religious affiliation, profession…, citizenship…, health and ability,… educational levels, and age". The collection of which the Spencer-Wood and Trunzo article is a part applies such ideas in a range of different contexts.

What emerges here is the issue of ethical behaviour, already noted (in previous chapters) as a running theme of this book. This has been dubbed as 'emancipatory' archaeology, defined by Saitta (2007, 267) as a practice "dedicated to expanding the conversation about what it means to be human [by fostering] critical thought about the determinants of *contemporary* lived experience [to bring about] positive social change". Such notions reflect the capacity to create presents and futures as discussed above, and will necessarily re-emerge in Chaps. 6 and 7—this is inevitably also a political and social archaeology—but the engagement of archaeologists with the wider public sphere is an essential component of this. An emancipatory archaeology is located somewhere near the 'democratic' extreme of the continuum of public engagement identified by Smith and Waterton (2009, 15–16). In its concern to make the world better, it is a concern to improve the condition of those with whom particular archaeologists are directly engaged. These are presented as case studies of practice in specific locations and contexts, especially those of conflict and risk for the communities involved, which also provide models for work elsewhere. Examples include:

- Saunders (2002) who charts the value of archaeology in identifying and documenting the evidence of human rights abuses, including State-ordered murders, in Guatemala and the Balkans;
- Crossland (2002) who reveals the complexity of issues over the exhumation and reburial of Argentina's 'disappeared';
- Ballbè et al. (2007) who seek to change the dominant discourse of silence over the Spanish Civil War and its consequences for the people of Catalonia;
- Ronayne (2007) who demonstrates the value of an alliance of archaeologists with the contemporary culture of people to be adversely affected by major infrastructural development projects;
- Meskell and van Damme (2008) who chart the role of archaeologists in obtaining recognition of living communities inhabiting areas subject to control for purposes of conservation;
- Kiddey and Schofield (2011) and Kiddey (2012) who demonstrate the therapeutic value of involving the homeless in archaeology; and
- Schaepe et al. (2017) who widen the therapeutic value to descendant communities.

While such an emancipatory archaeology has emerged only over past decades—from the 1990s to the early decades of the twenty-first century — the history of archaeology's engagement with a wider public goes back to the origins of the field. In late nineteenth century Britain, Augustus Henry Lane Fox Pitt-Rivers established his museum at Farnham to educate the workers on his estate into the inevitability of change through evolution rather than by revolutionary means, an aim (or at least an approach) that persists in the Pitt-Rivers Museum in Oxford (Hudson 1987). At this period—and indeed earlier—the distinction between a concern for the protection of ancient remains and providing public access to them did not yet exist: preservation inevitably allowed access, while the provision of public access justified efforts at preservation. John Jameson (2004) charts the rise of two distinct public archaeologies in the USA: from the pioneering work of Thomas Jefferson in promoting the heritage of Virginia directly to its population, through to the increasing involvement of State and Federal institutions in studying America's past, to a greater focus on preservation, and the ultimate dominance of preservation over interpretation in

Federal programmes (and see Chap. 5). He argues that it was theoretical debates about interpretation among archaeologists themselves—both academic and professional—that led to a revived interest in public engagement in the 1980s. Others (e.g. Frost 2004; Copeland 2004, 132–133) have outlined the range of public education programmes in archaeology that have developed in the USA and Britain over the past two decades, emphasising how recent the emergence of community projects has been. This development is also closely tied to an understanding of what archaeology is *for* and how—as mentioned earlier — it relates to the condition of being human: the latter is grounded in philosophical musings.

Archaeology as Philosophy

While archaeology is widely recognised as a philosophical discipline, archaeologists generally make poor philosophers. It is a philosophical discipline in three senses: that it derives—as do most modern areas of study—from the older, generalised 'natural philosophy' which combined interest in both the natural and the made in equal measure; because archaeologists are inspired by the work of philosophy to understand the world; and because archaeology is after all the study of how humans engage with the material world around them. Archaeologists, however, make poor philosophers because far too often they can be distracted from a direct concern with how humans and the material world interact to trying to divine what it is that a philosopher who they find interesting is really saying at a broader scale: they can become bogged down in the developmental flow of a particular philosopher's work rather than taking what is useful from another's ideas and ignoring what is not, the latter a process Olsen (2010; italics in original) describes as "*bricolage*" and one adopted elsewhere in my own work (Carman and Carman 2006, 223). The latter approach does not at all detract from the status of archaeology as a branch of philosophy but it is always essential to be clear what we are talking about.

Trigger emphasises that for scholars of the Middle Ages in Europe "no clear distinction was drawn between curiosities that were of natural and those that were of human origin" (Trigger 2006, 85): theirs was a philosophy based in religious belief grounded in the Biblical text and (to oversimplify) all things were perceived as the product of divine origin and therefore natural. Since the focus was upon identifying God's purpose—and there was expectation of an imminent end to all things worldly as prophesied—there was no need for any attempt at a systematic study of the past since the world was not deemed to be very old (only a few thousand years at most) and the full story was contained in religious text. For Trigger, the major shift in perceptions of the past comes with the so-called Enlightenment of the seventeenth and eighteenth centuries in Europe which allowed "a more naturalistic understanding of social processes with a firm belief in progress" (Trigger 2006, 99). This in turn led to the development of a series of tenets that were important for allowing the emergence of social sciences including archaeology:

- The 'psychic unity' of humanity which meant all were capable of contributing to and benefiting from advances in civilisation
- That "cultural progress is the dominant feature of human history"
- That cultural progress is not only technological but also moral and social, and that indeed they are causally linked with one another;
- That such progress leads to the perfection of human nature; and
- That progress is the result of rational thought (Trigger 2006, 100–102).

That this is a specifically Eurocentric view (as also acknowledged by Trigger) is itself significant: archaeology is the product of European invention and has been spread across the globe as a result of European colonisation and imperialism, as also emphasised in Chaps. 1 and 2. As a product of Western philosophy it is from such philosophies that archaeology draws, and it is to such philosophies that archaeology can make a contribution.

Thomas (2004) specifically links archaeology to the idea of modernity which he defines in terms of a number of connected characteristics (Table 3.1) with the effect that the modern condition is also characterised by "general unease and dissatisfaction" (Thomas 2004, 2). For Thomas, therefore, archaeology is a product of a particular historical condition reflecting a particular philosophical stance. His book charts the consequences of these for how archaeology has developed, especially in terms of a controlling discourse, with a view to the construction (as noted in Chap. 1) of a "counter-modern archaeology" whereby archaeologists abandon their authority over the treatment of others' pasts (Thomas 2004, 235–238). Mizoguchi

Table 3.1 Characteristics of modernity

	Archaeological authors	
Characteristics	Thomas 2004	Mizoguchi 2006
Capitalism	•	
Nation State	•	•
Industralisation	•	•
Improved communications	•	
Mercantilism	•	
Control of violence by the State	•	•
Surveillance	•	
Constant political struggle	•	•
Increasing urbanism	•	
Experience of agitation, turbulence and continuous change	•	•
Decline of tradition and religious belief	•	•
Erosion of established sources of social stability	•	•
Increasing self-interest and rationalism		•
Commoditisation		•
Individualism		•
Increasing differentiation		•

(2006) defines modernity in similar terms (Table 3.1) and especially in terms of its creation of "*functionally differentiated social systems*" and archaeology as "a communication system situated and reproduced" as part of that system (Mizoguchi 2006, 165, italics in original; see also Chap. 4) leading ultimately to increasing numbers of separate and discrete discourses. Focusing especially on issues around multiculturalism, he argues that as a result modernity is a period of crisis which needs to be overcome by the development of a "theory which constantly reminds us of the necessity of… communication across the boundaries of segmented" discourses (Mizoguchi 2006, 167). Koerner and Russell (2010) similarly see modernity as a period of crisis, because modernity's tendency to self-investigation produces a focus on the 'bads' rather than the 'goods' of the current state of things (Koerner 2010, 14). They argue that there are aspects of modernity hidden from view because of supposed "'settlements' of 'crises of representation'" whereby issues of trust reduce to expert testimony, that the current state of things requires entirely new responses, and which in turn deny the possibility of plural interpretations of pasts (Koerner 2010, 3). Instead, they offer a vision in which archaeologists "can provide and promote a diversity of knowledges – each of them developing their own relevance to the ways of addressing the contemporary world's needs and contemporary society's aspirations" (Koerner and Russell 2010, 415). All these commentators share a common aim—to construct an archaeology that contributes more than mere interpretations of the past and uses them to change the world.

While Thomas (2004), Mizoguchi (2006) and Koerner and Russell (2010) seek to overcome the limits of philosophical modernity for the benefit of the field and wider society, others—albeit part of the same wider project—look to specific philosophers to provide actual models for alternative ways forward. Holtorf and Karlsson (2000b) set out to challenge the dominant Anglophone discourse in archaeology especially in Europe; their contributors draw for inspiration upon a range of authors rarely encountered in English-language publications and the editors support their contention with a redrawn map of archaeological theory in Europe (Holtorf and Karlsson 2000a, 8). What is striking about the book is that few of the contributions consider philosophies of actual *materiality*—the focus is more often upon models of abstract concepts (a point made in a more general review by Olson 2010) than on understanding the physical *things* (objects, places, landscapes) that are the specific focus of archaeology. An exception is Karlsson (2000) who asks "why is there material culture rather than nothing?" but the discussion slips into a concern with distinguishing the ideas of Heidegger about 'Being' (a form of existence) and a 'being' (which exists), and so collapses to metaphysics that never actually answers the crucially important question. As Karlsson summarises unhelpfully: "I am, you are, the megalith in the illustration is and the illustration is, but why?" (Karlsson 2000, 80). In defence of the volume and the approach it takes, however, Kristiansen (2000, viii) points out that "the philosophical origins of theoretical thinking is what links archaeology to the humanities, and connects individual responsibility with the practical consequences of [archaeological] work…. [which inevitably] leads to a discussion of ethics".

Olson (2010), recognising and responding to the 'material turn' in the humanities (e.g. Finn 2018), argues that archaeology "is first and foremost a concern with things, and [therefore] archaeologists, as the most dedicated students of things, should naturally make their voices heard.... [because a] long held concern with things constitutes an intellectual skill that is clearly highly relevant" (Olson 2010). He criticises others—including those who claim a concern for materiality—for treating material things as epiphenomena of social and cultural practice, and therefore not real in themselves, effectively 'dematerialising' them. As he puts it, "Societies and cultures are otherwise rather consistently treated as collectives of humans held together by social relations and social forces—in short, by *people without things*" (Olson 2010; italics in original). Instead, he chooses to focus on the very materiality of things—including their simple characteristics as size, durability, persistence, and day-to-day function. These things contribute to their 'integrity' and their significance *in themselves*. As he puts it, things come to us "as things that work, not as symbolic consumables.... [The] tasks they fulfil are primarily due to their irreplaceability or superiority in conducting precisely these tasks and to their nonarbitrariness in significant operations" for which they were designed (Olson 2010).

The study of *how* things are made is especially the province of those who study technology and technique. While this is frequently done in the service of developing an archaeology of human cognition (e.g. Schlanger 1994) it too is a focus upon the tangible qualities of material and the physical processes by which it is turned into a thing. The idea of the *chaîne opératoire* can be—and has been—applied to the study of stone tool manufacture (Schlanger 1994; Bar-Yosef and van Peer 2009), rock art (Brady et al. 2022), metallurgy (Kuijpers 2018), ceramics (Roux 2019) and architecture (Pelmoine and Mayor 2020). It is a specifically materialist approach that distinguishes between discursive knowledge and practical knowledge, the latter being the most significant in making things and a combination of cognition and practical skill, especially the ability to 'feel' what is the right thing to do to achieve the desired and planned result. As an arena of practical knowledge it constitutes an ability to respond to changes in the material itself as the work of making proceeds, and adapt the cognitive plan to meet those changes. In this way an understanding of the maker's response to the quality of the material itself is ultimately a key component in the analysis of the *chaîne opératoire.*

It is this 'thingly' quality of things that is also emphasised in developing a 'symmetrical' archaeology (Witmore 2007) which ascribes agency to objects—not intentionality with which it is often confused (as pointed out by Lindstrøm 2015)—but their undeniable capability of affecting and acting on the world. Strang (2014, 141) explains the agency of water:

> Water has a multiplicity of effects upon people and environments, ranging from immediate molecular interactions in organisms; to carving out watercourses, depositing silt etc. in river valleys; to making vast hydrological movements and transformations at an atmospheric level. It similarly carves out a central space in the human imagination, which is predisposed to formulate concepts.

Other objects also 'do' things in this way, at the service of humanity, or to humanity's disservice, depending what they are. Such an approach takes us away from a concern with anthropocentrism and allows us to see the world in a new way, as composed of things that include the animate and the inanimate in equal measure. This is the basis of Olsen's (2010) argument that archaeology deserves a specific voice in the social scientific 'material turn' that directly concerns the essentially physical nature of a material world: rather than reducing to abstract conceptions the material things with which we engage minute by minute throughout our lives, we recognise them for what they do as independent entities.

The debate concerning how we should consider the 'things' with which we deal is a lively and subtle one that goes to the heart of how we are engaged with all the other things—objects, structures, places—with which we share the world. Lindström (2015, 229) "differentiates between effects and acts, effectants and actants" in her counter-argument to the symmetrical approach, limiting the capacity to *act* to "cerebral life forms" (Lindström 2015, 229) while at the same time accepting the idea of our necessary entanglement with everything else in the universe. Jones (2007, 223), by contrast, recognises that our entanglement also involves participation and it is this idea that drives his concern with constructions of memory. It is the positioning of things—including ourselves—in relation to other things that allows meanings to be constructed and thereby memory, which is created by "the continual and dynamic encounter between the subject and the material world he or she inhabits rather than an abstract… transaction between the external world and the mind" (Jones 2007, 26). Ultimately, he suggests that the items composing the material world should be treated as a series of indices, whereby it is in the links between things that our capacity to interpret and understand lies. Here, we do not treat things as epiphenomena of cerebral process nor as isolated entities 'in-themselves' but as components of networks and it is in the network that we find what we seek from the world.

Archaeologists conventionally present their interests in philosophy as a borrowing of others' ideas to suit archaeological purposes. This is the approach for the contributions to Holtorf and Karlsson (2000b) for instance, in which the products of the thoughts of chosen commentators are the start-point for their application to archaeological thinking. But as these writings—and those of Olsen (2010), Thomas (2004) and Jones (2007) among others—demonstrate, archaeologists are not slaves to the ideas of others. Instead, they are the source whereby archaeologists construct their own understanding of the world and their own discipline: as in most academic writing (if we are honest) archaeologists draw upon others not so much for inspiration but to support and help enunciate what they already know to be true. That is why the apprehension of the material world that is central to the archaeological product is at once distinctive and important. Other disciplines may claim to engage with the material, but this often collapses to seeing them as the epiphenomenon of culture that archaeologists can move beyond: we instead recognise them in terms of the 'thingness' they encompass and demonstrate.

As a realm of material engagement and production, and as outlined above, archaeology is capable of presenting itself in various guises: as a realm of expertise, as one of authority, as one of experience, as one of excitement, and as one of abstract cogitation about wider issues, all based on the direct engagement with material that is at the core of archaeological practice (Chap. 2). It is because of that basis in direct experience of the material that the oft-quoted (and challenged) distinction between theory and practice is seen to be irrelevant: to theorise (or engage with philosophical thinking) is an important archaeological practice in itself, which then feeds other aspects of practice. As all these different kinds of producers, archaeologists communicate with many kinds of 'others'—other categories of person, with each other, with abstractions, and with material things both animate and inanimate, present and gone. In doing so, they communicate both directly and indirectly, in both words and in actions, and in the way they organise themselves. This communicative aspect of archaeological practice is the topic of the next Chapter.

References

Alonso, F.G. 2013. Archaeology and Nationalism: The Development of Archaeology in Catalonia in the Early Twentieth Century. *Complutum* 24 (2): 131–144.

Ascherson, N. 2004. Archaeology and the British Media. In *Public Archaeology*, ed. N. Merriman, 145–158. London: Routledge.

Ballbè, E.G., J.O. Puigdoménech, E.S. Olives, and D.W. Steadman. 2007. The Archaeology of the Spanish Civil War: Recovering Memory and Historical Justice. In *Archaeology and Capitalism: From Ethics to Politics*, ed. Y. Hamilakis and P. Duke, 235–246. London: Routledge.

Barker, P. 1982. *Techniques of Archaeological Excavation*. London: B.T. Batsford.

Bar-Yosef, O., and P. van Peer. 2009. The *Chaîne Opératoire* Approach in Middle Paleolithic Archaeology. *Current Anthropology* 50 (1): 103–131.

Bland, R. 2004. The Treasure Act and the Portable Antiquities Scheme: A Case Study in Developing Public Archaeology. In *Public Archaeology*, ed. N. Merriman, 272–291. London: Routledge.

Borck, L. 2018. Constructing the Future History: Prefiguration as Historical Epistemology and the Chronopolitics of Archaeology. *Journal of Contemporary Archaeology* 5 (2): 213–302.

Börjesson, L. 2015. Grey Literature—Grey Sources? Nuancing the View on Professional Documentation. The Case of Swedish Archaeology. *Journal of Documentation* 71 (6): 1158–1182.

Brady, L.M., L. Taylor, S.K. May, and P.S.C. Taçon. 2022. Meaningful Choices and Relational Networks: Analysing Western Arnhem Land's Painted Hand Rock art Style Using *chaîne opératoire*. *Journal of Anthropological Archaeology* 65: 101396.

Brodie, N. 2012. Uncovering the Antiquities Market. In *The Oxford Handbook of Public Archaeology*, ed. R. Skeates, C. McDavid, and J. Carman, 230–252. Oxford: Oxford university Press.

Carman, J. 2015. *Archaeological Resource Management: An International Perspective*. Cambridge: Cambridge University Press.

Carman, J., and P. Carman. 2006. *Bloody Meadows: Investigating Landscapes of Battle*. Stroud: Sutton Publishing.

Chippindale, C. 1996. On writing About Archaeology in the English Language. *Archaeological Dialogues* 3 (1): 47–55.

Clarke, D.L. 1968. *Analytical Archaeology*. London: Methuen.

Cleere, H.F. 1984. World Cultural Resource Management: Problems and Perspectives. In *Approaches to the Archaeological Heritage*, ed. H.F. Cleere, 125–131. Cambridge: Cambridge University Press.

Colwell-Chanthaphonh, C., and T.J. Ferguson, eds. 2008. *Collaboration in Archaeological Practice: Engaging Descendant Communities*. Lanham: Altamira Press.

Connah, G. 2010. *Writing About Archaeology*. Cambridge: Cambridge University Press.

Cooper, M.A., A. Firth, J. Carman, and D. Wheatley, eds. 1995. *Managing Archaeology*. London: Routledge.

Copeland, T. 2004. Presenting Archaeology to the Public: Constructing Insights On-site. In *Public Archaeology*, ed. N. Merriman, 132–144. London: Routledge.

Crossland, Z. 2002. Violent Spaces: Conflict Over the Reappearance of Argentina's Disappeared. In *Matériel Culture: The Archaeology of Twentieth Century Conflict*, One World Archaeology 44, ed. J. Schofield, W.G. Johnson, and C.M. Beck, 115–131. London: Routledge.

Cuddy, T.W., and M.P. Leone. 2008. Understanding the Americanization of African descent groups through archaeology. In *Collaboration in Archaeological Practice: Engaging Descendant Communities*, ed. C. Colwell-Chanthaphonh and T.J. Ferguson, 203–223. Lanham MA: Altamira Press.

Daniel, G. 1975. *150 Years of Archaeology*. London: Duckworth.

Darvill, T. 2012. Archaeology as a Profession. In *The Oxford Handbook of Public Archaeology*, ed. R. Skeates, C. McDavid, and J. Carman, 373–394. Oxford: Oxford university Press.

Diaz-Andreu, M. 2007. *A World History of Nineteenth Century Archaeology: Nationalism, Colonialism and the Past*. Oxford: OUP.

Dietler, M. 1994. "Our Ancestors the Gauls": Archaeology, Ethnic Nationalism, and the Manipulation of Celtic Identity in Modern Europe. *American Anthropologist* 96 (3): 584–605.

Edgeworth, M. 2006. Acts of Discovery: An Ethnography of Archaeological Practice. In *British Archaeological Reports*, International Series 1131. Oxford: Archaeopress.

Ellesmere, Earl of. 1848. *Guide to Northern Archaeology*. London: Bain.

Enqvist, J. 2014. The New Heritage: A Missing Link Between Finnish Archaeology and Contemporary Society? *Fennoscandia archaeologica* XXXI: 101–123.

Everill, P. 2012. *The Invisible Diggers: A Study of British Commercial Archaeology*. 2nd revised ebook edition. Available at https://www.academia.edu/4244219/The_Invisible_Diggers_A_study_of_British_commercial_archaeology_2nd_Edition_. Accessed 7 Feb 2023.

Eze-Uzomaka, P.I. 2000. Museums, Archaeologists and Indigenous People : Archaeology and the Public in Nigeria. In *BAR International Series 904*. Oxford: Archaeopress.

Finn, M. 2018. Material Turns in British History: I. Loot. *Abstract: Transactions of the Royal Historical Society*: 285–232. https://doi.org/10.1017/S0080440118000026.

Frid, P., and C. Holtorf. 2021. *Wow! The Future is Calling!* Kalmar: Linnaeus University.

Frost, K.E.S. 2004. Archaeology and Public Education in North America: A View from the Beginning of the Millennium. In *Public Archaeology*, ed. N. Merriman, 59–84. London: Routledge.

Gänger, S. 2009. Conquering the Past: Post-War Archaeology and Nationalism in the Borderlands of Chile and Peru, c. 1880–1920. *Comparative Studies in Society and History* 51 (4): 691–714.

Gnecco, C., and P. Ayala, eds. 2011. *Indigenous Peoples and Archaeology in Latin America*. Walnut Creek: Left Coast Press.

Gramsch, A. 2000. 'Reflexiveness' in Archaeology, Nationalism, and Europeanism. *Archaeological Dialogues* 7 (1): 4–19.

Graves-Brown, P. 2007. Concrete Islands. In *Contemporary and Historical Archaeology in Theory: Papers from the 2003 and 2004 CHAT Conferences*, Studies in Contemporary and Historical Archaeology 4. British Archaeological Reports International Series 1677, ed. L. McAtatackney, M. Palus, and A. Piccini, 75–82. Oxford: Archaeopress.

Harlan, M.E. 2010. Black and White Literature, Grey Scholars. *Archaeologies* 6 (2): 270–288.

Harlin, E.-K. 2019. Sámi Archaeology and the Fear of Political Involvement: Finnish Archaeologists' Perspectives on Ethnicity and the Repatriation of Sámi Cultural Heritage. *Archaeologies* 15 (2): 254–284.

Harris, O. 2018. Archaeology is Process. *Archaeological Dialogues* 25 (1): 34–38.

Harrison, R. 2009. 180 Towards an Archaeology of the Welfare State in Britain, 1945–2009. *Archaeologies: Journal of the World Archaeological Congress* Special Issue "Archaeo-ethnography, Auto-archaeology: Introducing Archaeologies of the Contemporary past", 5(2), 238–262.

———. 2016. Archaeologies of Emergent Presents and Futures. *Historical Archaeology* 50 (3): 165–180.

Harrison, R., and J. Schofield. 2010. *After Modernity: Archaeological Approaches to the Contemporary Past*. Oxford: Oxford University Press.

Hodder, I. 1986. *Reading the Past: Critical Approaches to Interpretation in Archaeology*. Cambridge: Cambridge University Press.

———. 1989. Writing Archaeology: Site Reports in Context. *Antiquity* 63: 168–174.

Högberg, A., C. Holtorf, S. May, and G. Wollentz. 2017. No future in Archaeological Heritage Management? *World Archaeology* 49 (5): 639–647.

Holtorf, C. 2004. Incavation – Excavation – Exhibition. In *Material Engagements: Studies in Honour of Colin Renfrew*, ed. N. Brodie and C. Hills, 45–53. Cambridge: MacDonald Institute for Archaeological Research.

———. 2016. Pastness in Themed Environments. In *A Reader In Themed And Immersive Spaces*, ed. S.A. Lukas, 31–37. Pittsburgh: ETC Press.

———. 2017. Changing Concepts of Temporality in Cultural Heritage and Themed Environments. In *Time and Temporality in Theme Parks*, ed. F. Carlà-Uhink, F. Freitag, S. Mittermeier, and A. Swarz, 115–130. Hannover: Wehrhahn Verlag.

Holtorf, C., and H. Karlsson. 2000a. Changing Configurations of Archaeological Theory: An Introduction. In *Philosophy and Archaeological Practice: Perspectives for the 21st century*, ed. C. Holtorf and H. Karlsson, 1–11. Göteborg: Bricoleur Press.

———, eds. 2000b. *Philosophy and Archaeological Practice: Perspectives for the 21st Century*. Göteborg: Bricoleur Press.

Hudson, K. 1987. *Museums of Influence*. Cambridge: Cambridge University Press.

Jameson, J.H., Jr. 2004. Public Archaeology in the United States. In *Public Archaeology*, ed. N. Merriman, 21–58. London: Routledge.

Jones, A. 2007. *Memory and Material Culture*. Cambridge: Cambridge University Press.

Joyce, R. 2006. Writing Historical Archaeology. In *The Cambridge Companion to Historical Archaeology*, ed. D. Hicks and M. Beaudry, 48–65. Cambridge: Cambridge university Press.

Joyce, R.A., C. Guyer, M. Joyce, J. Lopiparo, and R. Preucel. 2002. *The Languages of Archaeology: Dialogue, Narrative, and Writing*. Oxford: Blackwell.

Karlsson, H. 2000. Why Is There Material Culture Rather Than Nothing? Heideggerian Thoughts and Archaeology. In *Philosophy and Archaeological Practice: Perspectives for the 21st Century*, ed. C. Holtorf and H. Karlsson, 69–86. Göteborg: Bricoleur Press.

Kehoe, A.B. 2012. Public Education in Archaeology in North America: The Long View. In *The Oxford Handbook of Public Archaeology*, ed. R. Skeates, C. McDavid, and C. Carman, 537–551. Oxford: Oxford University Press.

Kersel, M. 2012. The Value of a Looted Object: Stakeholder Perceptions in the Antiquities Trade. In *The Oxford Handbook of Public Archaeology*, ed. R. Skeates, C. McDavid, and J. Carman, 253–272. Oxford: Oxford University Press.

Kiddey, R. 2012. 'I's Never Thought About Me Being Part of the History: The Value of Heritage Work with Socially Excluded People. In *Integrating Archaeology: Science—Wish—Reality. International Conference on the Social Role, Possibilities and Perspectives of Classical Studies. Papers held in Frankfurt a. M. on 12–14 June 2012*, ed. N. Schücker, 185–190. Frankfurt a. M.: Römisch-Germanische Kommission.

Kiddey, R., and J. Schofield. 2011. Embrace the Margins: Adventures in Archaeology and Homelessness. *Public Archaeology* 10 (1): 4–22.

Koerner, S. 2010. Introduction: Rethinking 'Crises Over Representation'. In *Unquiet Pasts: Risk Society, Lived Cultural Heritage, Redesigning Reflexivity*, ed. S. Koerner and I. Russell, 1–7. Farnham: Ashgate.

Koerner, S., and I. Russell, eds. 2010. *Unquiet Pasts: Risk Society, Lived Cultural Heritage, Redesigning Reflexivity*. Farnham: Ashgate.

Kohl, P., and C. Fawcett, eds. 1995. *Nationalism, Politics and the Practice of Archaeology*. Cambridge: Cambridge University Press.

Kristiansen, K. 2000. Preface. In *Philosophy and Archaeological Practice: Perspectives for the 21st Century*, ed. C. Holtorf and H. Karlsson, vii–vii. Göteborg: Bricoleur Pressi.

———. 2012. Archaeological Communities and Languages. In *The Oxford Handbook of Public Archaeology*, ed. R. Skeates, C. McDavid, and C. Carman, 461–477. Oxford, Oxford University Press.

Kuijpers, M.H.G. 2018. A Sensory Update to the *Chaîne Opératoire* in Order to Study Skill: Perceptive Categories for Copper-Compositions in Archaeometallurgy. *Journal of Archaeological Method and Theory* 25: 863–891.

Lindström, T.C. 2015. Agency 'in itself'. A Discussion of Inanimate, Animal and Human Agency. *Archaeological Dialogues* 22 (2): 207–238.

Lucas, G. 2001. *Critical Approaches to Fieldwork: Contemporary and Historical Archaeological Practice*. London: Routledge.

Marshall, Y. 2002. What Is Community Archaeology? *World Archaeology issue on "Community Archaeology"* 34 (2): 211–219.

McDavid, C. 2004. Towards a More Democratic Archaeology? The Internet and Public Archaeological Practice. In *Public Archaeology*, ed. N. Merriman, 159–187. London: Routledge.

McGimsey, C.R. 1984. The Value of Archaeology. In *Ethics and Values in Archaeology*, ed. E.L. Green, 171–174. New York: Free Press.

McManamon, F.P. 1991. The Many Publics for Archaeology. *American Antiquity* 56: 121–130.

Merriman, N. 2004. Introduction: Diversity and Dissonance in Public Archaeology. In *Public Archaeology*, ed. N. Merriman, 1–18. London: Routledge.

Meskell, L., and L.S.M. van Damme. 2008. Heritage Ethics and Descendant Communities. In *Collaboration in Archaeological Practice: Engaging Descendant Communities*, ed. C. Colwell-Chthathaphonh and T.J. Ferguson, 131–150. Lanham: Altamira Press.

Mikhailov, D.A. 2013. Altai Nationalism and Archeology. *Anthropology & Archeology of Eurasia* 52 (2): 33–50.

Mizoguchi, K. 2006. *Archaeology, Society and Identity in Modern Japan*, Cambridge Studies in Archaeology. Cambridge: Cambridge University Press.

Morgan, C. 2014. Archaeology and the Moving Image. *Public Archaeology* 13 (4): 323–344.

Moshenska, G. 2017. Introduction: Public Archaeology as Practice and Scholarship Where Archaeology Meets the World. In *Key Concepts in Public Archaeology*, ed. G. Moshenska, 1–13. London: UCL Press.

Mrozowski, S.A. 2006. *The Archaeology of Class in Urban America*. Cambridge: Cambridge University Press.

Nilsson Stutz, L. 2018. A Future for Archaeology: In Defense of an Intellectually Engaged, Collaborative and Confident Archaeology. *Norwegian Archaeological Review* 51 (1–2): 48–56.

Olsen, B. 2010. *In Defense of Things : Archaeology and the Ontology of Objects*. Lanham: AltaMira Press. Available at ProQuest Ebook Central, https://ebookcentral.proquest.com/lib/bham/detail.action?docID=616256. Accessed 17 Mar 2023.

Pearson, M., and J. Thomas. 1994. *Theatre/Archaeology. The Drama Review* 38 (4): 133–161.

Pelmoine, T., and A. Mayor. 2020. Vernacular Architecture in Eastern Senegal: *Chaînes opératoires* and Technical Choices. *Journal of Material Culture* 25 (3): 348–378.

Petersson, B. 2017. Anachronism and Time Travel. In *The Archaeology of Time Travel: Experiencing the Past in the 21st Century*, ed. B. Petersson and C. Holtorf, 281–298. Oxford: Archaeopress Archaeology.

Petersson, B., and C. Holtorf, eds. 2017. *The Archaeology of Time Travel: Experiencing the Past in the 21st Century*. Oxford: Archaeopress Archaeology.

Praetzellis, A. 2003. *Dug to Death: A Tale of Archaeological Method and Mayhem*. Lanham: AltaMira press.

———. 2011. *Death By Theory: A Tale of Mystery and Archaeological Theory*. Lanham: AltaMira press.

Pratap, A. 2009. *Indigenous Archaeology in India: Prospects for an Archaeology of the Subaltern*, BAR International Series 1297. Oxford: Archaeopress.

Qin, D. 2004. The Effects of the Antiquities Market on Archaeological Development in China. In *Public Archaeology*, ed. N. Merriman, 292–299. London: Routledge.

Renfrew, C. 2000. *Loot, Legitimacy and Ownership*, Duckworth Debates in Archaeology. London: Duckworth.

Ronayne, M. 2007. The Culture of Caring and Its Destruction in the Middle East: Women's Work, Water, War and Archaeology. In *Archaeology and Capitalism: From Ethics to Politics*, ed. Y. Hamilakis and P. Duke, 247–266. London: Routledge.

Roth, B.J. 2010. An Academic Perspective on Grey Literature. *Archaeologies* 6 (2): 337–345.

Roux, V. 2019. The Ghassulian Ceramic Tradition: A Single *Chaîne Opératoire* Prevalent Throughout the Southern Levant. *Journal of Eastern Mediterranean Archaeology & Heritage Studies* 7 (1): 23–43.

Rubertone, P.E., ed. 2008. *Archaeologies of Placemaking: Monuments, Memories, and Engagement in Native North America*, One World Archaeology 59. Walnut Creek: West Coats Press.

Saitta, D.J. 2007. Ethics, Objectivity and Emancipatory Archaeology. In *Archaeology and Capitalism: From Ethics to Politics*, ed. Y. Hamilakis and P. Duke, 267–280. London: Routledge.

Saunders, R. 2002. Tell the Truth: The Archaeology of Human Rights Abuses in Guatemala and the Former Yugoslavia. In *Matériel Culture: The Archaeology of Twentieth Century Conflict*, One World Archaeology 44, ed. J. Schofield, W.G. Johnson, and C.M. Beck, 103–114. London: Routledge.

Schadla-Hall, T. 2004. The Comforts of Unreason: The Importance and Relevance of Alternative Archaeology. In *Public Archaeology*, ed. N. Merriman, 255–271. London: Routledge.

Schaepe, D.M., B. Angelbeck, D. Snook, and J.R. Welch. 2017. Archaeology as Therapy: Connecting Belongings, Knowledge, Time, Place, and Well-Being. *Current Anthropology* 58 (4): 502–533.

Schlanger, N. 1994. Mindful Technology: Unleashing the chaîne opératoire for an Archaeology of Mind. In *The Ancient Mind: Elements of Cognitive Archaeology*, ed. C. Renfrew and E.B.W. Zubrow, 143–151. Cambridge: Cambridge University Press.

Schlanger, N., L. Nespoulous, and J.-P. Demoule. 2016. Year 5 at Fukushima: A 'disaster-led' Archaeology of the Contemporary Future. *Antiquity* 90 (350): 409–424.

Schnapp, A. 1996. *The Discovery of the Past: The Origins of Archaeology*. London: British Museum Press.

Schofield, J., E. Praet, K.A. Townsend, and J. Vince. 2021. 'COVID Waste' and Social Media as Method: An Archaeology of Personal Protective Equipment and Its Contribution to Policy. *Antiquity* 95 (380): 435–449.

Schücker, N., ed. 2012. *Integrating Archaeology: Science—Wish—Reality. International Conference on the Social Role, Possibilities and Perspectives of Classical Studies*. Papers held in Frankfurt a. M. on 12–14 June 2012. Frankfurt a. M.: Römisch-Germanische Kommission.

Schülke, A. 2000. Archaeology and Art. In *Philosophy and Archaeological Practice: Perspectives for the 21st Century*, ed. C. Holtorf and H. Karlsson, 261–276. Göteborg: Bricoleur Press.

Seymour, D.J., ed. 2010. Special Issue: Black-and-White Issues About the Archaeological Grey Literature. *Archaeologies* 6 (2): 226–232.

Shanks, M. 1992. *Experiencing the Past: On the Character of Archaeology*. London: Routledge.

Shanks, M., and R.H. Mcguire. 1996. The craft of Archaeology. *American Antiquity* 61 (1): 75–88.

Shepherd, N. 2003. 'When the Hand That Holds the Trowel is Black ...' Disciplinary Practices of Self-representation and the Issue of 'Native' Labour in Archaeology. *Journal of Social Archaeology* 3 (3): 334–352.

Simonetti, C. 2018. *Sentient Conceptualisations: Feeling for Time in the Sciences of the Past*. London: Routledge.

Sinclair, A. 1989. Writing *ARC*-haeology. *Archaeological Review from Cambridge* 8 (2): 159–164.

Skeates, R., C. McDavid, and J. Carman, eds. 2012. *The Oxford Handbook of Public Archaeology*. Oxford: Oxford University Press.

Smith, L., and E. Waterton. 2009. *Heritage, Communities and Archaeology*, Duckworth Debates in Archaeology. London: Duckworth.

Smith, C., and H.M. Wobst. 2005. *Indigenous Archaeologies: Decolonising Theory and Practice*, One world Archaeology 47. Abingdon: Routledge.

Spector, J. 1991. What this awl Means: Towards a Feminist Archaeology. In *Engendering Archaeology: Women and Prehistory*, ed. J. Gero and M.W. Conkey, 388–406. Oxford: Blackwell.

Spencer-Wood, S.M., and M.C. Trunzo. 2022. Introduction to Archaeologies Special Issue on Intersectionality Theory and Research in Historical Archaeology. *Archaeologies* 18 (1) Special Issue: Intersectionality Theory and Research in Historical Archaeology: 1–44.

Strang, V. 2014. Fluid Consistencies. Material Relationality in Human Engagements with Water. *Archaeological Dialogues* 21 (2): 133–150.

Thomas, J. 2004. *Archaeology and Modernity*. London: Routledge.

Thomas, S. 2012. Archaeologists and Metal-Detector Users in England and Wales: Past, Present, and Future. In *The Oxford Handbook of Public Archaeology*, ed. R. Skeates, C. McDavid, and C. Carman, 60–81. Oxford, Oxford University Press.

Thurley, S. 2013. *Men from the Ministry: How Britain Saved Its Heritage*. New Haven/London: Yale University Press.

Trigger, B.G. 2006. *A History of Archaeological Thought*. Cambridge: Cambridge University Press.

Tzanidaki, J.D. 2000. Rome, Maastricht and Amsterdam: The Common European Heritage. *Archaeological Dialogues* 7 (1): 20–33.

Van Sebille, E., Delandmeter, P., Schofield, J., Hardesty, D., Jones, J. and Donnelly, A. 2019. Basin-scale Sources and Pathways of Microplastic that Ends Up in the Galápagos Archipelago. Open Science Discussions. https://doi.org/10.5194/os-2019-37. Accessed 13 Mar 2023.

Varouchakis, V. 2017. Indigenous Archaeologies of Crete, 1878–1913. *Public Archaeology* 16 (1): 42–66.

Willems, W.J.H. 2000. Discussion: Archaeology and Europe: Reflexiveness' and Action. *Archaeological Dialogues* 7 (1): 36–45.

Witmore, C.L. 2007. Symmetrical Archaeology: Excerpts of a Manifesto. *World Archaeology* 39: 546–562.

Zena, A.G. 2018. Archaeology, Politics and Nationalism in Nineteenth- and Early Twentieth-Century Ethiopia: The Use of Archaeology to Consolidate Monarchical Power. *Azania: Archaeological Research in Africa* 53 (3): 398–416.

Chapter 4
Archaeology as Communication

As mentioned in Chap. 3, Mizoguchi (2006) defines archaeology as "a communication system situated and reproduced" as part of a system of *"functionally differentiated social systems"* located in the current fragmented historical context we call modernity (Mizoguchi 2006, 165, italics in original). As such, archaeology is one communication system among many others from which archaeology can be distinguished and with which it interacts. Such a definition emphasises some of those aspects of archaeology discussed in previous chapters of this book and to be discussed in those to follow: and especially the various ways in which archaeologists 'speak' (using the word in its broadest sense) to each other and the rest of the world. A brief introduction to this was provided in Chap. 3 but here we shall investigate in more depth and more specifically by examining the types of communication indulged in by us and to whom they are directed. It will also necessarily engage a concern for the reception of such communications.

Mizoguchi (2006) locates communication at the core of his understanding of archaeology in modernity because, as he puts it, "sociality emerges when two or more individuals are co-present… in a locale" and "communication is *the* basic unit of sociality" (Mizoguchi 2006, 37–38). For archaeologists, he argues, the medium of communication is archaeological material—somewhat akin to Harris' (1973) notion of 'currency' in group psychology, the medium of 'currency' being the most important aspect ordering relations between group members. The currency is what unites the group but it is group unity—as reflected in the idea of the group as a group—which is the ultimate aim of group solidarity. For Mizoguchi, the crucial group membership so far as archaeology is concerned is that of the nation state, for which archaeology provides a historical foundation and also a clear referent in terms of extent and boundaries, and therefore also the context within which archaeology is possible. Mizoguchi is particularly concerned with the role of archaeology in Japan but it is a model that can be applied elsewhere because across the globe archaeology and the creation of nation-states are inextricably linked, as copious publications have demonstrated (e.g. Kohl and Fawcett 1995; Diaz-Andreu and

J. Carman, *Archaeological Practice as Politics and Ethics*,
https://doi.org/10.1007/978-3-031-69828-6_4

Champion 1996; Diaz-Andreu 2007). The connection between archaeology and the 'imagined community' of the nation (Anderson 2016) establishes a single discourse which allows intra- and extra-disciplinary communication which can be presented as value-free and scientific.

As Mizoguchi goes on to explain, however, the rise of multivocality presents a problem for archaeology as a scientific, apparently value-free unitary ethno-national discourse. Accepting that others have an equal right to interpretations of the past—interpretations that derive from sources other than material or other evidence and are fundamentally different from those of the archaeological discipline—is open to the critique that it privileges certain stances and views over others—such as that of Indigenous voices over those of the academically trained (Mizoguchi 2006, 124). A recognition that others have an equal legitimacy to talk of pasts in fundamentally different ways, as Mizoguchi argues, it leads to the conclusion that "*no unified/definite knowledge of the world is possible*" (Mizoguchi 2006, 130, italics in original) and, rather than opening up a wider scope for debates between alternatives, has actually created "*mutually segregated*, even sometimes hostile, discursive spaces/fields which do not communicate with one another" (Mizoguchi 2006, 130, italics in original). In particular, he suggests, we have seen within archaeology the rise of "an ever-increasing number of self-claimed *new* approaches/stances" and therefore the formation of small circles of colleagues who share similar approaches to their work but separate themselves from others who take a different approach (Mizoguchi 2006, 131, italics in original; and see Carman 2017 for similar ideas as well as their development in Chap. 7 of this book).

Mizoguchi's solution to this 'crisis'—as he terms it (Mizoguchi 2006, 121)—is constant and continuing evaluation of what archaeologists say to create a "continuously rearticulated archaeological dialogue" (Mizoguchi 2006, 163). This idea is very similar to that of others—such as McDavid's (2000) emphasis on the value of ongoing 'conversations' with others, regardless of whether they build consensus, which nonetheless provide an opportunity to "make our own truth claims as loudly and forcefully as we can" (McDavid 2000, 231). Those like McDavid see this as a means to create what Mizoguchi would also like to create—"a more democratic archaeological praxis [involving] free and open encounters between each other" (McDavid 2000, 231–232). Whether our actual communications serve this purpose or impede it will be examined in what follows.

Within the Field

As we all know, archaeologists communicate to each other in diverse ways. In publications, any of the various and quite different types of written product may appear as stand-alone pamphlets, as articles in journals, or as parts or wholes of books: there are no strict rules about which is most appropriate for a particular mode of communication. And we also speak verbally to each other directly, individually and collectively, especially at conferences where we inform others of our current work in the field. Partly in jest—but only partly—I have sometimes referred to the most appropriate collective noun for archaeologists as 'an argument'; but disputing with one another is so often what we do in our communications. Paradoxically, out of those disputes arise the community that is archaeology (Carman 2017).

As discussed in Chap. 3, the 'classic' form of archaeological output is the formal report deriving from fieldwork, often described as 'grey' literature (Seymour 2010) because of its ambiguous status as published material. There is a standard structure to such reports, whether of fieldwalking, survey or excavation, or indeed other types of work such as desk-top survey, conveniently summarised by Clarke and Popescu (2014):

- A non-technical summary (the whole report in miniature)
- An introduction setting out all preliminaries, why the work was conducted and by and for whom
- Geological and topographical setting of the location of the work
- Archaeological and historical background to the location of work
- Aims of the work
- Methodology
- Results—a detailed description of everything found, the contexts for such finds, and summaries of objects retrieved
- Discussion of the results, outlining how they fit within existing patterns of knowledge, and an overall Conclusion
- Supporting data in the form of tables and charts, plans and drawings
- Illustrations demonstrating finds, contexts and the conduct of the work
- A list of other literature (grey or otherwise) referenced in the report

This is material specifically aimed at other professional archaeologists and unlikely to be read by anyone else—and certainly not in full, although some others who are interested may look at the non-technical summary and any final statements in the Discussion section. As Hodder (1989) and others (Lucas 2001) have pointed out, in modern reports of this kind information is so often transmitted in very abstract terms, referring to specific locations by coded 'context number', soil type and colour by reference to a numbered Munsell Chart, and clearance of soil and retrieval of objects as if carried out independently of human action. Nonetheless, however unreadable, such reports are the fundamental building-blocks of archaeological interpretation and knowledge. They present data which is open to re-interpretation by others and incorporation into their own work: it is archaeologists speaking to other archaeologists in 'reality' mode, saying in effect, this is how it is.

However, lest it be thought that such work is forever to be relegated to the dusty archive from which it only rarely escapes, reports on work can also be published in other forms—especially as journal papers by the major period or area societies, such as the *Proceedings of the Prehistoric Society* or the *African Archaeology Journal;* or as monographs which will have a wider readership. The latter form of output is, as may be expected, generally reserved for work considered to have a wider significance to the field than the standard site report which frequently will effectively replicate work elsewhere while nonetheless adding to the knowledge-base with new data. Publication as a monograph in particular will involve more than mere description: there will be a stronger engagement with theoretical and methodological issues than in the conventional report, and will aim to produce results of importance to archaeologists in conducting their own work. Schliemann's work in the nineteenth century at Hissarlik and elsewhere, published in two volumes (Schliemann 1976a, b) is of this form, but also includes significant extraneous material we rarely find today, such as the author's autobiography (Schliemann 1976a, 1–67) and extensive Prefaces by distinguished academics (Sayce 1976; Virchow 1976) designed to promote the author's status in the eyes of the world. As Lucas (2001, 19) points out about Schliemann's status today, however, "We may deride Schliemann as a treasure hunter, even see some hypocrisy in his allusion to the greed of the workmen and his own secretive removal of the finds [in an extract quoted by Lucas], but he was working in a very different intellectual climate and it is too easy to dismiss his archaeology as a primitive version of today's progressive techniques".

More recent researchers also publish their fieldwork as monographs in much the same format as Schliemann but without so much obvious self-serving content: I choose two at random. Barrett et al.'s (1991) volume on work in the 1970s and 1980s at Cranborne Chase, England, is structured very much as a conventional site report although its purpose—to convince others of the usefulness of a particular approach to monuments and landscape in order to elucidate aspects of social process in the past—is made very explicitly early on (Barrett et al. 1991, 6–8). After this we are introduced to the specifics of the archaeological investigation in terms of what it tells us about the area under study through the prism of landscape development and change through recognised chronologies based upon the developed 'Three Age' system. The book ends with a discussion of the latest period they cover—that immediately preceding the Roman invasion of Britain—emphasising the value of others' prior work and hoping theirs makes a useful contribution: there is no final summary or conclusion returning us to the theoretical stance outlined in the opening pages. The book is also at pains to give credit to the work of others on the project and indeed who included text on aspects of the work in the final publication. It contrasts interestingly with Schliemann's efforts at self-promotion, however necessary he may have felt the need in his own time.

The second modern random example is Foard and Curry's (2022) work on the site of the Bosworth battlefield, also in England. This differs from the Cranborne Chase work in that—although equally collaborative—the main collaboration is between an archaeologist (Foard) and a historian (Curry): as a result, the book is structured slightly differently from both the site report and Barrett et al. (1991) despite containing very similar kinds of information. A crucial factor in the difference is that the Bosworth book is first of all about the discovery of the location of an event rather than an attempt to understand the distribution of and relations between objects in a landscape. Like the other, however, it serves to justify a particular approach to the object of research: in particular, the methods of battlefield archaeology using landscape reconstruction and survey, especially by metal detector. The opening chapters are—like Schliemann's (1976a, b) and Barrett et al.'s

(1991)—introductory but longer since they concern not only the purpose of the work and prior work to find the battlefield, but also the historical background to the events of the battle and what we know of late medieval warfare in Britain. Landscape reconstruction—a combination of historical and archaeological knowledge derived largely from map work—follows, and then the results of survey. A chapter specifically devoted to interpretation of the finds is succeeded by one devoted to understanding the evidence for gunpowder weapons and what this adds to our understanding of warfare in the period, and then a discussion offering 'a new perspective on the battle' leading to a short conclusion again justifying the approach. The work as a whole thereby offers a template and model as guidance for other work to identify and understand particular types of landscape.

All these works—Schliemann's (1976a, b), Barrett et al.'s (1991) and Foard and Curry (2022), as well as many others (see e.g. Hamilakis 2013; Harrison and Schofield 2010; McGuire and Paynter 1991; Bapty and Yates 1990; Clarke 1968; Sabloff 2008; Little 2007, all of whom cite one or more examples of specific archaeological projects)—thus seek to influence future research rather than, as more common for site reports or collections thereof, simply to present data that may contribute to understanding pasts. The primary audience for each of them—despite commercial publication and general sale—is the academic and practicing archaeological community rather than a wider general readership. The data they offer is for a purpose other than speaking for itself as an offering for alternative interpretations. Works of this kind are archaeologists speaking to other archaeologists in 'persuasive' mode—saying, in effect, this is how you should be.

As we have seen in previous chapters, and indeed as this book proceeds to do, archaeologists write to each other not only about the sites and material they investigate, but also about the nature of the work they undertake. There is an underlying persuasiveness about much of this work, but it also goes beyond setting out the best way of achieving certain kinds of results by asking fundamental questions about what it is we do, why in that way and indeed why at all. Previous chapters have introduced us to the work of Lucas (2001), Thomas (2004), and this chapter to that of Mizoguchi (2006), all of whom have sought to understand the nature of archaeological work and to offer a way forward from the way things are done at present. Each of these offers a means whereby archaeology—or aspects of the field—can be located within their historical and cultural context and each of them identifies structural issues at the heart of the discipline deriving from its position in modernity and argue for changes in approach needed to address them. Structurally they share some similarities so that Lucas (2001) and Thomas (2004) order their chapters by relevant topics but do not refer to specific investigations of sites except in passing; their approach is generally what would be called 'theoretical' because lacking data in the form of material retrieved from sites. Mizoguchi (2006) is equally theoretical in his approach, but in focusing upon the Japanese experience he draws much more upon the evidence of Japanese pasts, including the kinds of site plans and illustrations more typical of the site report (Mizoguchi 2006, 55–117).

Works such as site reports and those from Lucas (2001), Thomas (2004) and Mizoguchi (2006)—as well as others of similar kind—can be described as 'unidirectional' in that they consist of one voice speaking to whoever wishes to listen. Apart from peer review in advance of publication (on which see more below) there is no space for others' commentary or rejoinder. The same is true of most work published as articles in journals, but some provide for more genuine conversations on topics. The journal *Archaeological Dialogues* specialises in such exchanges, in which an article is offered for commentary by others selected by the journal editor with a final response by the original author, but other journals such as *Archaeologies*, *Norwegian Archaeological Review* and the *Journal of Community Archaeology and Heritage* also provide space for such conversations (see e.g. Conkey 2005; Moshenska 2008; Carman et al. 2009; Carman 2017). One example is Nativ (2018) on what it is that archaeologists study with responses from others with similar concerns but differing perspectives. Nativ closes his response to commentary and the discussion — reflecting my own preferred term for the collective noun for archaeologists and the preferences of both Mizoguchi (2006, 163) and McDavid (2000, 231)—with the "hope [that] the somewhat provisional remarks offered here will help keep this discussion going" (Nativ 2018, 42).

Such commentary on what it is that archaeology is—and like this book—is neither the statement of bald facts (called above the 'reality mode') nor an attempt to direct others in a particular direction (called above the 'persuasive' mode) but something else: this is archaeologists asking questions about what it is that they do and debating the available answers. It is best described as archaeologists in 'critical reflexive' mode: there is a sense in which the answer matters less than the fact of the inquiry in the first place. It is an exercise in serious reflection—echoing archaeology's ability to contribute to philosophical musings discussed in Chap. 3—that prevents the ossification of the discipline and its reduction to the production of irrelevant information about dead pasts: it leads us to maintain our relevance and therefore importance in the contemporary world in which we work.

There are of course publications other than those discussed above including for instance works on the history of the discipline (e.g. Trigger 2006; Murray and Evans 2008; Stout 2008), biographies and celebrations of key figures (e.g. Patton 2007; Bánffy 2022) and interviews with current major contributors (e.g. Rathje et al. 2013; Ribeiro 2023), among others, many of which are also directed outside the professional and academic discipline. There is one key distinction between some publications and others that needs to be made and recognised. So-called 'grey' literature is rarely subject to review for its accuracy or significance, except by those in whose name it is produced, which is one reason its 'publication' status remains ambiguous, as discussed above and in Chap. 3. In order to achieve full publication status in a journal or as a book, other works, including those discussed above, will have undergone the process called 'peer review' whereby the text goes—frequently anonymously—to others to assess its potential significance and appropriateness for release to its intended audience. Reviewers will submit reports to the editors of journals or the publishers of books to indicate any problems they may have with seeing the work in print or any improvements or clarifications they believe should be made. Quite often, despite the chagrin the author may feel on having their carefully constructed work criticised, such advice is helpful and does improve the work and is therefore useful. An argument against it is that it may impede the publication of truly innovative work because it falls outside the norms of publication content or style; and the process has been charged with bias against certain categories of person and indeed frequent poor performance by reviewers. Nonetheless, the process is designed to ensure that only work truly worthy of release to the disciplinary audience achieves publication. Any work not subject to peer review can legitimately be considered as of lower quality, less reliability and inferior significance.

One form of communication that is rarely truly peer reviewed because of the difficulties involved in doing so but is never challenged as a result is that in which one archaeologist speaks directly to others. This is most usually in the context of the academic conference where archaeologists sharing common interests gather to exchange news and present the results of their work. Many conferences are narrow in their concerns—in particular areas of the world, specific periods of the past, particular types of material or specific approaches—and sometimes held once only, but others can be more general and regional in extent or global and held at regular intervals. Nonetheless, the sharing of common interest applies here too: those who attend the annual conferences of the Society for American Archaeology or the European Association of Archaeologists have specific interests in archaeology in those territories; while those who attend the 4-yearly meetings of the World Archaeological Congress subscribe to the principles and ideals of that organisation.

The academic conference is a very particular kind of event and includes more than the formal presentation of research results, even though that is the usual justification, especially for attracting funding to attend such events. As Deardorff (2015, 316), writing from the perspective of political science, notes, "Academics go to conferences for a variety of reasons, many of which are determined by the stage of their career or the type of institution that employs them": but the same is true of any discipline including archaeology. Among the reasons for attending any particular conference are:

- To disseminate one's work and obtain recognition
- To promote oneself to others in your field
- To discover others in your area of work with whom you may wish to collaborate, or against whom you may need to compete (or indeed both depending on when you next encounter one another)
- To reassure oneself that one's interests are not unique
- To seek out potential examiners (for graduate degrees) or reviewers of one's work
- To meet old friends and colleagues
- To arrange formal meetings with colleagues, possible future employers or publishers
- To find out about new developments in one's field
- To socialise with others in one's field
- To attend social events organised as part of the activities
- To visit new places (especially for conferences held as part of a regular series that may be located in a different place each time)
- To usefully spend research funds given by one's institution.

Any of these—or any number of them—is a perfectly valid reason for conferencing and they indicate some of the other things that happen at conferences other than presentations of papers in formal sessions. Large conferences—especially globally prestigious ones such as those of the International Union of Pre- and Proto-historic Societies or the World Archaeological Congress, but also those of regional significance such as the Society for American Archaeology or the European Association of Archaeologists—may well begin with an elaborate opening ceremony with speeches and some kind of performance and an equally intricate closing ceremony to thank all for attending and wish everyone safe journeys, as well as plenaries to discuss and agree matters of ethics or policy. Many of any kind—even those of relatively short duration and limited attendance—may also include less formal sessions—such as open discussion 'ethics bowls' to discuss difficult decisions all may face, field trips to locations of interest to participants in the surrounding region, a formal conference dinner, and perhaps a party, all designed to allow participants to meet and talk informally.

The informal socialising is required for two main reasons: to provide a necessary break from the formal rules of the academic session; and more importantly to assert the unity in endeavour of all those attending the conference. The conference is thus a device for creating a disciplinary community that recognises itself as a community. The rules in place for the formal 'scientific' sessions are unspoken but can be quite rigid. They include requirements on speakers—that they attend the session, remain throughout and deliver a paper to a set limit on time that meets the requisite formal standard of delivery—and on the audience that they do not interrupt speakers, leave during talks or otherwise interfere with the conduct of the session, and applaud at the close of each talk: the mood is very much that of the classroom, to which even the most senior attendees must conform as if a student. Questioning of individual speakers is to be expected, if not required, but also follows a set of unspoken conventional rules, breaches of which (such as lack of courtesy, irrelevance to the topic of the presentation, or failure to listen to the response even if it is a non-answer to the question as put which is sometimes the case) will be frowned upon by those present. It is generally considered proper to allow more scope to junior scholars than those more experienced in formal questioning, but otherwise the space of the conference is one of equality, so that even the most junior of attendees is free to approach even the most eminent and to talk. This too is an aspect of the community-building function of the conference.

The outcomes of conferences vary. For some it is merely the fact that they took place and provided such opportunities for individuals to meet as were available. For others there may be some formal publication as an output, usually a collection of papers derived from those as presented (e.g. Mathers et al. 2005), sometimes the text of those as presented (e.g. Carver 2004), sometimes—especially in the case of very large and diverse events—only short abstracts of papers presented (e.g. AIA 2019). Other outcomes may be less visible: the confidence given to a junior scholar after their first presentation; the career progression of another; the formation of a new research collaboration; the inspiration of a new project by what others said in their various papers; or simply new friendships spawned in the context of social events.

The rise of the online and hybrid conference—where attendance may be either in person or online—has necessarily adversely affected the community-building and -maintenance function of these events. It necessarily places the focus of participation onto the formal presentations and sessions, especially the former since online questioning or commentary on contributions is limited to use of the 'chat' function of whatever platform is being used to allow remote participation. This in turn gives greater authority to the person running the session who acts as mediator between commentator and presenter: the scope for supplementary or follow-up questions is very limited indeed if they need to be typed for delivery. Overarching this, it is impossible to take part in social events remotely, and so the opportunity to speak directly to one another in informal settings entirely removed. It is true that remote participation offers certain advantages: it is cheaper since there is no need to pay for transport to or accommodation during the course of the event, and so benefits those with less access to resources; it saves time, since it may require a week away from home to attend a conference which is spread over 3 days but may in fact only occupy a half day, one full day and another half day, so perhaps 16 h in all; and it may allow recording and replay of key presentations and therefore the opportunity to consider in some detail the content of papers given. On the other hand, all can be mitigated by other means: more resources may be provided to those attending such events; greater recognition given for the choice to do so and additional support provided for such absences, perhaps also providing funding to support the attendance of colleagues, students or partners; and any conference may include recording of presentations to be made available to all attendees on demand. Nonetheless, the appeal of online attendance for funding bodies and academic institutions is great since it saves them money and may be considered 'greener' due to the reduced travel involved, even though organising such events in that way harms the structure of those disciplines the communitarian spirit of the conference should be promoting.

If the function of communication—in all its diverse forms—with those like oneself is to create sociality then that is what communication within the discipline of archaeology seeks to achieve. Whether in 'persuasive', 'reality' or 'critical reflexive' mode, in writing or in speech, archaeologists identify themselves to and with other archaeologists by those communications, and allow debate and argument within the field. Even though Mizoguchi (2006) recognises the rise of separate discourses within the field, these are nonetheless all contained within the field itself and thus create a single, albeit fragmented, discourse of archaeology to which all archaeologists can subscribe. The latter is Mizoguchi's (2006, 165, italics in original) "communication system situated and reproduced" as part of a system of *"functionally differentiated social systems"*, those other functionally differentiated systems operating outside and alongside that of archaeology. From the 'outside' archaeology thus appears as a single mode of representation, although from the inside it appears fragmented and divided, making it sensible to consider internal modes of communication as separate from those directed outside the discipline, which is the topic of the next section.

Before moving on however it is worth mentioning some forms of writing not covered above, of which discussion is reserved to the following. These in particular are those publications aimed at students of the discipline, especially the textbook and the encyclopaedia. These can be considered as hybrid or ambiguous in terms of communication within the discipline: students are not yet full members of the field and those of us so established have little need for them as part of our regular reading. Indeed, no student further into their studies than the first year of a degree course—and no-one studying at the postgraduate level—should need recourse to them because they operate at too general a level. They do have some value, however, in outlining the basics of aspects of archaeology with which readers may be unfamiliar. It is that lack of familiarity that provides readers with their motive for accessing such material, and therefore how they can be classed as aimed beyond the field.

Beyond the Field

The single albeit fragmented discourse identified above is the way in which archaeologists converse with others, in equally diverse ways as those to others within the discipline. As discussed in Chap. 3, Connah (2010) in particular has outlined some of the types of publications involved while others (e.g. Potter 1994; Swain 2007; Grima 2017; Copeland 2004) discuss the other forms of communication that may be adopted. Not all—or perhaps even many—are written: it is more likely to take the form of presentations of material, either on-site or in various kinds of off-site display since archaeological fieldwork and artefactual analysis are the stereotypical visions of archaeology so often presented. Nonetheless, to those who wish to enter the field—but are so often not considered truly part of it—will engage with other images we like to present: especially works on theory and those exposing the many sub-fields archaeology encompasses, as well as guidance on possible career paths.

Text books take a number of forms. There are very general such texts (e.g Renfrew and Bahn 2016; Greene 2010, both of which are works periodically updated and revised; Ollich-Castanyer 2012) and others that specialise on aspects of archaeology: of regions and periods (e.g. Phillipson 2005; Piggott 1965; Milisauskas 2002; Megaw and Simpson 1979; Coles and Harding 1979) or indeed the globe (Clark 1969; Barsky 2022); of archaeological theory (e.g. Johnson 1999; Hodder 2001); of fieldwork and other archaeological investigative practice (e.g. Drewitt 2011; Hodder 1989; Roskams 2001; Richards and Britton 2020); of the development of the discipline (Trigger 2006); of public archaeology (Moshenska 2017c; Skeates et al. 2012; Thomas and Lea 2014); or of archaeological heritage management (e.g. Carman 2002; Carman 2015). Some such works are sole- or co-authored while others are edited collections; some remain unaltered even though passing into the out-of-date as the discipline develops in new directions, while others are frequently updated into new editions; some, even though unaltered to reflect newer ideas, will remain as classics in the field to be consulted by generation after generation. It is unlikely many of these will be read by general readers although some may, but in general they will find their place on the reading lists of those in formal study, especially at university level. They are not usually the kind of text that can be classed as 'critical reflexive' but they are not really representations of an objective material reality nor a deliberate attempt at persuasion to a particular point of view: although all will have been written from a particular perspective, the aim is less to encourage attachment to a particular viewpoint than to attract to membership of the community of archaeologists. They can therefore perhaps be classed as archaeologists writing in 'educative' mode, with one aim being to encourage readers to develop their own understandings as their knowledge and skill level grows.

The 'educational model' of communication is one of three for archaeology identified by Holtorf (2007, 150), the others being 'public relations' and 'democratic'—all are aimed at the non-professional public. The educational model reflects Merriman's (2004), 5–6) 'deficit' characterisation of the public whereby they are deficient in knowledge and understanding and therefore in need of enlightenment. The 'public relations' model reflects the arguments of e.g. Cleere (1984) and McGimsey (1984) for engagements with the public for the benefit of archaeology. The democratic model reflects the ideals of Thomas' (2004) 'counter-modern' archaeology, McDavid's (2000) call for 'conversations' and the involvement of others in collaborative enterprises (e.g. Colwell-Chanthaphonh and Ferguson 2008a) rather than mere 'outreach'. Where Holtorf's 'educational' model differs from my 'educative' mode—apart from the intended recipients—is that his is actually a version of what he otherwise calls 'public relations': the intention of education of the wider public who do not wish to become archaeologists is inevitably to offer an understanding of the field that they can be interested in and support. Students in formal education do not require to be persuaded of the value of the field since they have already volunteered to join (even though many may choose not to continue in an archaeological career): the aim of reaching out to those outside the field is instead to justify our existence and the maintenance of the supply of resources—economic, political, etc.—to allow us to continue in our work.

Holtorf's (2007) paper is one of several contained in an issue of the *European Journal of Archaeology*—the journal of the European Association of Archaeologists—dedicated to discussions of communication in archaeology. While some of the authors focus on communication between archaeologists (e.g. Venclová 2007), the majority consider communications with others. Levy (2007) in particular addresses communication with the kinds of 'stakeholders' to be found in relation to archaeology in North America, among whom are primarily "Native Americans…, but other stakeholders also make claims for the priority of their interests because of land ownership, long-time occupation, [or] commitment to allegedly broader issues such as economic development or environmental protection" (Levy 2007, 174; and compare with McManamon 1991 referenced in Chap. 3). She emphasises that in trying to communicate outside the field, "we are not in a position to control the message… [and are] also working against some entrenched public impressions of the field" (Levy 2007, 172). Her recommendations directed at the EAA concern the establishment of working groups on "public outreach, modern archaeological education, working with volunteers, working with local communities and on writing for the mass media" to create "coalitions [of interest] that reach across the multiple communities with an interest in the past" and serve the purpose of "the conservation of archaeological sites" (Levy 2007, 180), again emphasising how efforts in archaeological education of the wider community are ultimately designed to serve not those communities but the benefit of archaeology and archaeologists.

Others take a different approach, although generally agreeing that effective communication will serve the interests of the discipline. In particular, Watkins (2006) is keen to emphasise the significance of word-use by archaeologists in their communications outside the discipline. As an example (Watkins 2006, 101 and 103–106), he points to the implications inherent in describing a people's non-continuation of occupation at a place with use of the term 'abandonment', which implies a relinquishing of any claim of association with the place: in the case of Indigenous peoples such 'abandonment' may mean nothing of the kind as they remember the place as one of significance even if they can no longer live there. In similar vein, he considers the language used in debates about the peopling of North America which collapse to rival claims of control and authority to no-one's benefit. As he puts it, "Archaeologists agree that humans came into the North American continent from some other location, but the use of contemporary geopolitical terms in oversimplified statements such as 'Europeans [or Africans or Asians] were the first Americans' runs the risk of creating significant misunderstanding of the complexities of data interpretation involved" (Watkins 2006, 112).

The educative mode is most obviously applied in the context of formal education and training, but this too can have implications for wider community engagement. Prescott (2013, 43) argues that since "Communication is based on knowledge, and knowledge is based on the insights training engenders" so that "a revamping of university programmes in archaeology is perhaps a starting point". In this vein, as mentioned in Chap. 1, Kersel (2007) addresses the debates concerning the repatriation of materials to countries of origin in her 'Great Debate' exercise, bringing "to the foreground often-overlooked aspects of the ramifications of archaeological practice" (Kersel 2007, 63) especially as they relate to claims of ownership by the nation state. In the same volume, Smith and Burke (2007) offer a complex exercise concerning the equally complex kinship structures of Indigenous Australian societies, with a view to "counter aspects of injurious stereotypes that… underpin… perceptions of Indigenous peoples…. that remove Indigenous peoples from both the present and from the arena of sociopolitical action" (Smith and Burke 2007, 97–99). Taking such an approach one stage further, Quave et al. (2021) designed a programme for undergraduate students in the US specifically to "center [*sic*] the needs and interests of marginalized peoples and archaeological stakeholders" to "reveal the limitations and possibilities of archaeological knowledge production, encouraging [participants] to ask, 'Who benefits from archaeology, who is harmed, and how might we imagine a different way of studying and engaging with the past?'" (Quave et al. 2021, 87). One noticeable result was a change in the demographics of enrolment on the course and others on archaeology (Quave et al. 2021, 95–96)—with a marked increase in the proportion of non-White students. Quave et al. (2021, 97) argue that their purpose was "to reconstruct the processes of the past in service of a better future—and to do so in a way that centers [sic] the needs and priorities of diverse stakeholders [and that those] values should be incorporated into teaching the discipline from the very first day students encounter it". By means such as this, the communication of archaeology not only extends archaeological discourse but also serves the idea of archaeology as a realm of political action, a theme to be developed in Chap. 6.

Making our use of language more approachable—as urged by Watkins (2006) and Connah (2010)—is one part of the story to aid our communication with others. The rise of digital resources presents opportunities and challenges of its own, as is evident from contributions to Hageneuer's (2020) collection which covers the range of different uses for digital tools: as games to reveal the processes involved in experiencing archaeology, in presenting materials in museums, and as a tool in formal education. Bonacchi (2017) distinguishes between 'broadcasting' and 'participatory' modes of digital engagement in public archaeology, and the value of mining media use data for understanding the attitudes and behaviours of those accessing archaeology by those means. But of course—as Bonacchi's (2017) recognition of a 'participatory' mode for digital engagement recognises—there is more to the digital environment than writing: it also acts as an arena of experience, albeit one separate from actual external reality; it offers the opportunity to *do* things as well as read things, and advanced developments in areas such as haptics and virtual reality allow users to touch the otherwise untouchable and otherwise immerse themselves in a different environment. This acts as a convenient link to the other ways in which archaeology communicates to the world—in terms of its tangible physical practices and its direct engagement with the material world.

The multiple usages of the word 'archaeology' referred to at the outset of Chap. 1 are those that come into play here. Archaeology as a set of things people called archaeologists *do* (the main focus of this book) is one such usage; archaeology as the *material* to which archaeologists do things is quite another. In terms of engaging with those who are not themselves archaeologists, the former involves encouraging those others to do as we do—by getting them to undertake forms of archaeological investigation; the latter involves the presentation to them of the physical materials that work produces—either directly or via other media. Both can be considered as examples of the 'educative' and 'public relations' mode of activity identified by Holtorf (2007) but the direct involvement of others in actual archaeological investigation also merges into his 'democratic' model. Fieldwork as a form of training exercise as discussed in Chap. 2 also reappears as an aspect of this since encouraging others to work with us, or indeed for us, requires that they be reasonably proficient in the necessary skills. Presenting the material we uncover is less democratic than educative by its very nature—presenting anything is a one-way process—but it also requires skill and knowledge to understand and appreciate what is being presented (as recognised by e.g. Merriman 1991; Walsh 1992) and so there is a necessary additional educative dimension involved.

As others (e.g. Smith and Waterton 2009, 15–16) have recognised, and is discussed in Chap. 3, there is a 'sliding scale' of involving others in archaeological work, from mere outreach—demonstrating to others what we do (Moshenska 2017b, 5) — through forms of consultation with stakeholders, to actual collaboration with interested communities, and finally to giving others control over the work itself (see also Thomas and Lea 2014). Consultation is a much-abused term, frequently covering nothing but mere outreach—whereby a community of stakeholders are told what is happening with limited opportunity to comment—but it also extends through to advice in advance and an opportunity for genuine dialogue (although it may not actually change anything)—to a full opportunity for a community to modify and become involved in the work itself. Requirements that consultation with interested parties take place in relation to archaeological work—as for instance in US legislation—may not specify exactly what that consultation should involve, but where it does it may nonetheless be clear that ultimate authority lies with the archaeologist rather than the community being consulted: consultees are then in a purely advisory role, whose advice may be ignored (Bendremer and Richman 2006, 108). Going further than consultation leads to various forms of collaboration. Colwell-Chanthaphonh and Ferguson (2008b, 10–11) describe it as a continuum: from a form of participation where those involved bring their separate needs and where compromises ensure mutual agreement over which needs shall be met; to full collaboration where aims and goals are agreed mutually by all those involved from the outset to ensure all needs are equally met. A truly 'democratic' approach goes further yet than collaboration: it privileges the opinions, views and needs of the non-archaeologist over those of the archaeologist, who then acts as the servant rather than the partner of the stakeholder community (e.g. McDavid 2004).

An unspoken—and frequently ignored—complication to developing a truly democratic or 'open' (Moshenska 2017b, 9) archaeological practice is that of the issue of ownership of the process and what it produces. This is not a technical legal concern with intellectual property rights (although it affects them too) but rather an interest in addressing the question of whose archaeology it is (see e.g. Carman 2005; Anderson and Geismar 2017). To complicate things further, the term as applied here covers both meanings: archaeology as a set of practices, and the material with which archaeologists engage. The point is that if archaeology truly belongs to the communities who lay claim to it—as local or descendant communities, and especially among the latter Indigenous groups—then a democratic archaeology will give that community total control over the material and any use to which it is put, including that of scientific research; but that will then effectively exclude others from appropriating it (see also discussions in Scarre and Coningham 2013). Models of alternatives other than exclusive proprietary property rights (as discussed in Carman 2005) include such non-exclusive means such as that of 'cognitive ownership' (Boyd et al. 2005) which denies no-one access but also serves to prevent others from appropriation. The notion of cognitive ownership goes beyond the recognition of particular stakeholders to the grant of rights to anyone who wishes to claim association. The problem with such an approach—although it has a libertarian egalitarian appeal—is that it potentially extends rights of claim beyond the local or descendant community who may legitimately be able to claim that the archaeology is the product of their biological or cultural ancestors to the exclusion of others.

Despite this complication, however, the principle of the recognition of the claims of others to their pasts over that of archaeologists is the dominant theme and the overriding issue.

In discussions of giving others access to archaeological sites, there is a frequent confusion over what we mean. It often collapses to the mere presentation of already-excavated sites or upstanding remains (see e.g. Grima 2017; Gustaffson and Karlsson 2012) rather than the presentation of archaeologists at work, although allowing public access to working sites is not unusual. Such visits inevitably involve some one-way, 'top-down' styles of presentation, especially an introductory presentation to explain what visitors will encounter, such as how changes in soil colour and texture indicate the presence of an archaeological feature, as well as some idea of what those working on site may be doing if not actually digging, such as recording sections or discussion with a colleague how to proceed. Visitors will usually then be guided around the site—or maybe wander around by themselves, following marked routes—to talk to excavators who will explain what they are doing at that moment. The visit may then finish with an introduction to the finds hut to obtain some idea of finds processing as well as being shown some of the more interesting finds so far (see e.g. Carman 2002, 131–132). This is clearly not anything collaborative, since the visitors are merely passive recipients of what they are allowed to experience, but it is a chance to see archaeology as an active productive process (see Chap. 3) rather than a concern for static remains. There is an element of exploitation of the excavators, however, who are put on display in the same way as objects in a museum (or upstanding monuments in a landscape) and treated as the objects of visitors' gaze rather than active agents in their own right.

Allowing a public to gaze over archaeology—in any sense of the word, as practice or thing—is ultimately the most common engagement and form of communication with others we undertake. Whether the ruins of monumental structures, the objects in a museum or excavators in a trench, archaeology is presented to others as an exotic departure from their everyday experience. The process of visiting a site during its excavation is, for the visitor, not unlike that of visiting a factory or looking through gaps in the hoarding around a building site: it is an examination of others at work, which will be different from one's own; it will be a form of theatrical experience where actors perform for the viewer's benefit. For the excavators, who like factory workers but not necessarily building workers will know they are objects of gaze, it will be experienced as being under observation by those who may wish to make judgements upon the work itself, the manner in which it is conducted, aspects of dress or hairstyle or other aspects of their appearance: such comments may be made quietly or in plan hearing since visitors may imagine some kind of screen between them and the objects of their view. Ideally, excavators will choose to volunteer for such scrutiny but it may be difficult to refuse requests from a paying employer. Of course, upstanding remains, ruins and other static objects have no such choice but they too are objects of the gaze and places of experience.

There are two ways of presenting archaeological sites in the sense of the physical remains of large-scale structures. One is to present them as they are—as ruins, as incomplete, as excavated from the soil—albeit with some measure of consolidation to prevent damage to structure or visitors, as well as various interpretive devices to allow what Grima (2017, 82–90) calls 'intellectual access' by visitors. The other is in the form of reconstructions as they once were, or may have been (e.g. Mizoguchi 2006, 1–13). The latter is also a form of interpretation and, as Mizoguchi (2006, 5–7) explains, tends to give the impression on any such site that excavation revealed, as he puts it (Mizoguchi 2006, 4), that "everything [at the site] happened at once" rather than—as in Mizoguchi's example of a Yayoi period settlement — being a gradual process over a long period of time with multiple phases of occupation of varying styles and density. This impression of completeness—that sites in the past appeared fully formed and final in a single instant—is what showing people around sites during excavation (or, in the case of upstanding structures, during restoration and preparation) may be intended to achieve, but that assumes repeated visits by the same people to experience how these places change as work continues. It is not clear to what extent visitors who come once may do so again—and indeed the experience of organised visits to such sites is that once is very often enough. The impression of instantaneous, compressed activity in pasts is inevitably one we communicate whether we wish to or not.

Reconstructed and carefully conserved sites tend to be those which—one way or another—represent an image of their host country as those making decisions about its past would like to present. Staying in Japan, Hudson (2021) argues that a Japanese ideology of 'eco-nationalism' draws especially upon the archaeology of the Jōmon period to support an idea of the Japanese as both ethnically different from neighbouring populations as well as the West, and, in addition and crucially, as morally superior to both. Inaba (2012) makes the same argument about the Japanese approach to cultural landscapes, although from a less critical position. Similarly, Mizoguchi (2006) goes beyond his one example to outline the construction of a distinctive Japanese national identity based upon interpretations of the Jōmon period and succeeding Yayoi and Kofūn periods based around the particular material cultures of each. In such a formulation, the pottery of the Jōmon (14,000 to 1000 BCE) and the 'keyhole' tombs of the Kofūn (300 to c.600 CE) which are described in terms of Imperial burial rites, each serves to locate 'traditional' aspects of Japanese ethnicity and culture in the deep past. But of course the same applies elsewhere: in England, we carefully preserve and present medieval castles and later country houses; in Italy, Roman sites; in Greece, Classical and Byzantine structures; in Egypt, those of the Pharaonic past and mosques; other possible pasts on which to focus are often less clearly evident. These 'national' pasts are not ones of long-term change and process, but of sudden lurches from one stability to another which we help to construct even as we deny and often attempt to subvert the idea. In part it is an inevitable legacy of the division of pasts into convenient periods for our own purposes (e.g. the 'Three-Age System' and its derivatives: Chap. 3), in which change apparently only occurs between periods (or so it must seem to those others who follow our work).

Despite these unintended consequences, our communications by way of presenting the material with which we engage are meant to serve a range of different purposes. Since the presentation of sites is essentially no different from that of the presentation of individual objects — both are of static things to a passive audience — a perspective from the museum is not out of place here (and indeed quite often included in such discussions: e.g. Moussouri 2014; and the many references in Moshenska 2017c and Skeates et al. 2012). Writers on museum display (e.g. Belcher 1991; Dean 1996) offer a 'sliding scale' of exhibition style. For Belcher (1991) they fall into two main types, being either object or visitor centred, and representing various aims and purposes depending on the expected response of the viewer (Table 4.1).

A systematic display places similar objects in relation to one another, either in terms of development or chronological order. A thematic display will place objects together as assemblages, as in terms of associated function or ritual purpose. A participatory display (akin to Bonacchi's (2017) use of the term as mentioned above) involves visitors in some kind of action—inviting them to handle or manipulate objects. In all such cases—as the generic name of this kind of display indicates—the focus is upon the object rather than the visitor. Indeed, the visitor is considered of less significance than what is on display: there is a real sense in which the display serves the interests of the objects by presenting them in a very particular manner (and see also the discussion of the agency of objects in Chap. 3). The visitor then also serves the interest of the object on display by responding in an appropriate manner—usually passively, but in the case of a participatory display only in the terms allowed by the display. Most if not all archaeological sites and monuments are presented in exactly this manner: visitors are allowed to engage with the site by walking around it, clambering over parts of it, touching it, and so on, but not living in it or engaging in other activities associated with its past usages except as allowed by those responsible for the site. Even those dedicated to meaningful engagements with our publics nonetheless also have concerns for the preservation of archaeological resources (e.g. McCarthy and Wickert 2022).

Belcher's (1991) visitor centred displays present a continuum from visitors as passive recipients of information to having some measure of control over their experience. The educational style delivers information—by signboards and notices, guided routes, or guided tours—designed to teach the visitor certain things but only certain things. The exhibition planned as entertainment may also educate (it may be described as 'edutainment') but is primarily concerned to offer the visitor an enjoyable experience. Both treat the visitor as passive recipients of what is delivered to them. The emotive display expects a response from the visitor, either in terms of an appreciation of the aesthetic experience or some other kind of response (e.g. of

Table 4.1 Types of display. (From Belcher 1991)

Object centred display	Visitor centred display
Systematic	Didactic / educational
Thematic	Entertaining
Participatory	Emotive—Aesthetic
	Emotive—Evocative / romantic
	Responsive to viewer's presence
	Interactive with viewer's wishes

'dark' heritage—Uzzell 1989) but it remains a one-way process from presenter to responder. The responsive display works the other way: the display notices the presence of the visitor and then does something—lights come on or go off, a door opens or shuts, it shows a video or produces a sound—but it is purely programmed response in which the role of the visitor is limited to triggering the event. The interactive display sits at the far end of the continuum, allowing the visitor a high degree of control over the display by inviting them to take part in moulding the experience to their wishes, highly akin to Bonacchi's (2017) use of the term in relation to digital resources. Nonetheless, none of these represent any kind of collaborative effort—the display, object- or visitor-centred—is organised by and for the purposes of those controlling the display. In terms of archaeological sites and monuments, any of these styles may be applied—or indeed more than one at any such site. Despite Belcher's separation of the styles into a continuum, any of the emotive, responsive or interactive styles may ultimately serve the purpose of education or entertainment, or both, confirming the overarching control retained by those managing the site. All such presentations and provision for access by others thereby also serve the interests of archaeology and confirm the authority of the discipline over its material. Any such presentation is therefore delivered in the 'educative' mode discussed above.

All of the above forms of communication—in written texts, in images, in presenting sites, and in tours of active excavations—are aimed at those who are actively interested in archaeology, and can be classed as 'allies' of the discipline (as discussed in Chaps. 3 and 7). These we address in educative mode but also as collaborators and—albeit more rarely—as those to whom we grant the right to control the work we do on their behalf. There are others to whom we also communicate, although these we consider more as threats and even as enemies and so the form of communication takes a different form. The practice of promulgating laws to protect archaeological sites—thereby giving authority to those who controlled them—was established in the nineteenth century and continued into the 20th and 21st (Carman in press): the message here was of course about the legitimacy of such control and the illegitimacy of any other claim that might be made over the material. While a 'democratic' form of engagement with others is increasingly desired by us, the grant of rights over archaeological material can be effected by law—as in the case of the US *Native American Graves Protection and Repatriation Act* of 1990 (Chari and Lavallee 2013)—but also serves to define and limit who has any such right and the circumstances in which it applies. Other efforts elsewhere to encourage others who retrieve the kinds of material of interest to us from the soil—such as metal detectorists and treasure hunters (e.g. Thomas and Stone 2009; Thomas 2012) — are also directed only at those who follow prescribed practices. The discovery that some such persons are of value in archaeological work—collectors in identifying objects, or metal detectorists in discovering battlefield sites—may also encourage archaeologists to develop relationships with them. Such endeavours effectively move particular types of person from one category—threat or enemy—to another, such as ally or collaborator, but only under the terms set by us.

Others we seek to communicate with—but generally only in educative mode—are those who have very different understandings of pasts and indeed of archaeology. Schadla-Hall (2004) and Moshenska (2017a)—among others—argue for some recognition as potential allies of those who subscribe to 'alternative archaeologies' such as the existence of 'lost lands' such as Atlantis or Mu, the construction of monuments by visiting aliens from outer space, or the misidentification of geological features as archaeological structures, as well as the 'finds' of remains of Noah's Ark or other objects described in ancient and religious texts. While Schadla-Hall (2004, 271) argues for a clear distinction to be made between some kinds of alternative pasts and others—and for some to be strongly challenged because of the threat they pose to reasoned understandings of the world, but others to be "acknowledged and celebrated as elements in the diverse ways in which people experience the past" — Moshenska (2017a, 137) calls for their study from "sociological, anthropological, historical and philosophical perspectives" because, like it or not, "they are very much part of archaeology". Nonetheless, we class those who do not subscribe to an archaeological view of what constitutes material culture and how it came to be, of what is real and what is not, as outside the realm of the field: we can recognise that they see the world differently and treat them as objects of investigation, but ultimately they lie outside the scope of our remit. We can communicate to them, but their return communications generally remain unheard in the discourse that is archaeology.

Of course those who sit outside the field have their view of us, and it is always interesting—and sometimes chastening—to obtain an understanding of how others see us. Holtorf in particular has examined archaeology from the perspective of popular culture (see e.g. Holtorf 2004, 2005) and identified the various visions the wider non-archaeological public have of us:

- That we seek treasures from under the ground
- That we do so in exotic locations
- That like detectives we piece together the past from its fragmentary remains (Holtorf 2004, 43–45).

All these have some truth, but as this book (hopefully) demonstrates, and archaeologists know, there is much more to the work than looking for 'treasures' and indeed digging, that we do much more work in mundane locations—especially our own countries and indeed towns and cities—than distant lands, and while using evidence to reconstruct pasts is what we do it is a collective exercise rather than one involving one omniscient investigator *a lá* Sherlock Holmes or Hercule Poirot. More recent work has approached the same topic in a different manner: the NEARCH project (Kajda et al. 2017) investigated public attitudes towards the socio-cultural value of archaeology across European states. An overarching result was that "archaeology is seen as socially valuable and that most Europeans maintain positive attitudes towards its development and investment" (Kajda et al. 2017, 14) and that it can contribute much towards addressing the world's current problems. While greater engagement with publics was desired because "democratization of knowledge is an important aspect… because archaeologists, and the ethics frameworks in which they work are obligated to share this knowledge as part of the public good", there was also strong belief that archaeology is primarily a responsibility of the State and to meet its obligations to the public on whose behalf it works, archaeologists need to broaden the scope of their concerns. In general, then, the public—in Europe at least—not only have a positive regard for the field but a reasonable understanding of its place in modern society and share archaeologists' visions for its future.

It is noticeable that while there are parallels for the internally-oriented 'reality' and 'persuasive' modes of communication for the external audience in the 'educational' and 'public relations' modes, there is no parallel for the 'critical reflexive' mode. The 'democratic' forms of engagement are not really designed to be 'critical reflexive' but are essentially a modified form of 'reality' mode since they treat the communities with which we engage as equals in a common endeavour. Even though forms of public archaeology such as that in Annapolis (Potter 1994; Carman 2002, 133–135) may take a critical perspective on the history of sites, they are not in themselves critical approaches to the work of archaeology itself. As a result, so far as engagements with others are concerned, archaeology is presented as a unified set of practices lying outside those others' normal experience which we do not invite them to critique or investigate for themselves. It follows that democratic modes of archaeological practice are not as fully democratic as we might like or may like to claim: ultimately all our efforts in communicating, collaborating and working with others are designed to serve the purposes of archaeology, even though there may be other benefits that accrue to those others such as widening their own understanding of the world. As the realm of communication identified by Mizoguchi (2006), archaeology—like all academic fields—is not especially democratic so far as those outside the discipline are concerned, but within it remains a realm of diverse voices capable of genuine critical reflection on its own practices.

When communicating with ourselves in 'critical reflexive' mode we are quite capable of recognising the faults in our own claims to authority and power and asserting our ideal of a democratic approach to studying the past. In some of our practice—especially in various forms of collaborative and democratic endeavour—we can put those ideals into operation in our engagements with others. But ultimately we use our authoritative status as experts to guard the entrance to our discipline: others must learn from us as to how to make sense of pasts from its material remains or be treated by us as outsiders. That claim to authority rests firmly upon the role of archaeology as an arm of public service, and especially as Mizoguchi (2006) and others (e.g. Kohl and Fawcett 1995; Diaz-Andreu and Champion 1996; Diaz-Andreu 2007) note, as an agent of the state. It is as public service that we examine the field in the next chapter.

References

Anderson, B. 2016. Imagined Communities: Reflections on the Origin and Spread of Nationalism. Revised edition. London: Verso.

Anderson, J., and H. Geismar. 2017. *The Routledge Companion to Cultural Property*. London: Routledge.

Archaeological Institute of America. 2019. *121st Annual Meeting Abstracts Volume 43*. Boston: Archaeological Institute of America.

Bánffy, E. 2022. Marija Gimbutas 100: A 'Grande Dame' of Twentieth-Century Prehistory and Her Connections with Hungarian Archaeology. *Historical Studies on Central Europe* 2 (2): 305–315.

Bapty, I., and T. Yates, eds. 1990. *Archaeology After Structuralism: Post-structuralism and the Practice of Archaeology*. London: Routledge.

Barrett, J., R. Bradley, and M. Green. 1991. *Landscape, Monuments and Society: The Prehistory of Cranborne Chase*. Cambridge: Cambridge university Press.

Barsky, D. 2022. *Human Prehistory: Exploring the Past to Understand the Future*. Cambridge: Cambridge University Press.

Belcher, M. 1991. *Exhibitions in Museums*. Leicester: Leicester University Press.

Bendremer, J.C., and K.A. Richman. 2006. Human Subjects Review and Archaeology: A View from Indian Country. In *The Ethics of Archaeology: Philosophical Perspectives on Archaeological Practice*, ed. C. Scarre and G. Scarre, 97–114. Cambridge: Cambridge University Press.

Bonacchi, C. 2017. Digital Media in Public Archaeology. In *Key Concepts in Public Archaeology*, ed. G. In Moshenska, 60–72. London: UCL Press.

Boyd, W.E., M.M. Cotter, J. Gardiner, and G. Taylor. 2005. Rigidity and a Changing Order... Disorder, Degeneracy and Daemonic Repetition: Fluidity of Cultural Values and Cultural Heritage Management. In *Heritage of Value, Archaeology of Renown: Reshaping Archaeological Assessment and Significance*, ed. C. Mathers, T. Darvill, and B.J. Little, 43–57. Gainseville: University of Florida Press.

Carman, J. 2002. *Archaeology and Heritage: An Introduction*. London: Continuum.

———. 2005. *Against Cultural Property: Archaeology, Heritage and Ownership. Duckworth Debates in Archaeology*. London: Duckworth.

———. 2015. *Archaeological Resource Management: An International Perspective*. Cambridge: Cambridge University Press.

———. 2017. Herding Cats: Building Archaeological Communities with Responses by Kristiansen, K. Pyburn, K. A., Henson, D. *Journal of Community Archaeology and Heritage* 4(2), 1–62.

Carman, J. in press. 'Governed by Legislation': What Laws Do to Archaeology. In Diaz-Andreu, M. & Coltofean, L. (eds) *Oxford Handbook on the History of Archaeology*. Oxford University Press.

Carman, J., N. Anfinset, and G. Moshenska. 2009. *Norwegian Archaeological Review* 42 (2): 173–182.

Carver, G., ed. 2004. *Digging in the Dirt: Excavation in a New Millennium*, British Archaeological Reports International Series. Vol. 1256. Oxford: BAR.

Chari, S., and J.M.N. Lavallee. 2013. *Accomplishing NAGPRA : Perspectives on the Intent, Impact, and Future of the Native American Graves Protection and Repatriation Act*. Corvallis: Oregon State University Press.

Clark, G. 1969. World Prehistory: A New Outline. 2nd edition of *World Prehistory*. Cambridge: Cambridge University Press.

Clarke, D.L. 1968. *Analytical Archaeology*. London: Methuen.

Clarke, R., and L. Popescu. 2014. *An Introduction to Archaeological Report Writing. Jigsaw Cambridgeshire Best Practice users' Guide*. Bar Hill: Jigsaw Cambridgeshire.

Cleere, H.F. 1984. World Cultural Resource Management: Problems and Perspectives. In *Approaches to the Archaeological Heritage*, ed. H.F. Cleere, 125–131. Cambridge: Cambridge University Press.

Coles, J.M., and A.F. Harding. 1979. *The Bronze Age in Europe: An Introduction to the Prehistory of Europe, c. 2000–700 BC*. London: Methuen.

Colwell-Chanthaphonh, C., and T.J. Ferguson, eds. 2008a. *Collaboration in Archaeological Practice: Engaging Descendant Communities*. Lanham: Altamira Press.

———. 2008b. Introduction: The Collaborative Continuum. In *Collaboration in Archaeological Practice: Engaging Descendant Communities*, ed. C. Colwell-Chanthaphonh and T.J. Ferguson, 1–32. Lanham: Altamira Press.

Conkey, M.W. 2005. Dwelling at the Margins, Action at the Intersection? Feminist and INDIGENOUS ARCHAEOLOGIES with Comments by Smith, L., Lippert, D, Million, T. Korstanje, A. and Blau, S. *Archaeologies* 1: 9–80.

Connah, G. 2010. *Writing About Archaeology*. Cambridge: Cambridge University Press.

Copeland, T. 2004. Presenting Archaeology to the Public: Constructing Insights On-site. In *Public Archaeology*, ed. N. Merriman, 132–144. London: Routledge.

Dean, D. 1996. *Museum Exhibition: Theory and Practice*. London: Routledge.

Deardorff, M.D. 2015. Reconsidering the Scholarly Conference for the Contemporary Academic. *PS: Political Science and Politics* 48 (2): 315–318.

Diaz-Andreu, M. 2007. *A World History of Nineteenth Century Archaeology: Nationalism, Colonialism and the Past*. Oxford: OUP.

Diaz-Andreu, M., and T. Champion, eds. 1996. *Nationalism and Archaeology in Europe*. London: Routledge.

Drewitt, P.L. 2011. *Field archaeology: An Introduction*. 2nd ed. London: Routledge.

Foard, G., and A. Curry. 2022. *Bosworth 1485: A Battlefield Rediscovered*. Barnsley: Oxbow Books.

Greene, K. 2010. *Archaeology: An Introduction*. 5th ed. London: Routledge.

Grima, R. 2017. Presenting Archaeological Sites to the Public. In *Key Concepts in Public Archaeology*, ed. G. Moshenska, 73–92. London: UCL Press.

Gustaffson, A., and H. Karlsson. 2012. 'Changing of the Guards': The ethics of Public Interpretation of Cultural Heritage Sites. In *The Oxford Handbook of Public Archaeology*, ed. R. Skeates, C. McDavid, and J. Carman, 478–495. Oxford: Oxford university Press.

Hageneuer, S., ed. 2020. *Communicating the Past in the Digital Age: Proceedings of the International Conference on Digital Methods in Teaching and Learning in Archaeology (12–13 October 2018)*. London: Ubiquity Press.

Hamilakis, Y. 2013. *Archaeology and the Senses: Human Experience, Memory, and Affect*. Cambridge: Cambridge University Press.

Harris, T.A. 1973. *I'm OK—You're OK*. London: Pan Books.

Harrison, R., and J. Schofield. 2010. *After Modernity: Archaeological Approaches to the Contemporary Past*. Oxford: Oxford University Press.

Hodder, I. 1989. Writing Archaeology: Site Reports in Context. *Antiquity* 63: 168–174.

———., ed. 2001. *Archaeological Theory Today*. Cambridge: Polity Press.

Holtorf, C. 2004. Doing Archaeology in Popular Culture. In *The Interplay of Past and Present*, 42–49. Södertörns högskola.

———. 2005. *From Stonehenge to Las Vegas: Archaeology as Popular Culture*. Walnut Creek: Left Coast Press.

———. 2007. Can You Hear Me at the Back? Archaeology, Communication and Society. *European Journal of Archaeology* 10 (2–3): 149–165.

Hudson, M.J. 2021. *Conjuring Up Prehistory: Landscape and the Archaic in Japanese Nationalism*. Oxford: Archaeopress.

Inaba, N. 2012. Cultural landscapes in Japan: A Century of Concept Development and Management Challenges. In *Managing Cultural Landscapes. Key Issues in Cultural Heritage*, ed. K. Taylor and J.L. Lennon, 109–129. London: Routledge.

Johnson, M. 1999. *Archaeological Theory: An Introduction*. Oxford: Blackwell.

Kajda, K., A. Marx, H. Wright, J. Richards, A. Marciniak, K.S. Rossenback, M. Pawleta, M. van den Vries, K. Boom, M.P. Guermandi, F. Criado-Boado, D. Barreiro, A. Synnestvedt, K. Kotsakis, K. Kasvikis, E. Theodoroudi, F. Lüth, M. Issa, and I. Frase. 2017. Archaeology, Heritage, and Social Value: Public Perspectives on European Archaeology. *European Journal of Archaeology* 21 (1): 96–117.

Kersel, M. 2007. The Great Debate: Archaeology, Repatriation and Nationalism. In *Archaeology to Delight and Instruct: Active Learning in the University Classroom*, ed. H. Burke and C. Smith, 55–65. Walnut Creek: Left Coast Press.

Kohl, P.L., and C. Fawcett, eds. 1995. *Nationalism, Politics and the Practice of Archaeology*. Cambridge: Cambridge University Press.

Levy, J.E. 2007. Archaeology, Communication, and Multiple Stakeholders: From the Other Side of the Big Pond. *European Journal of Archaeology* 10 (2–3): 167–184.

Little, B.J. 2007. *Historical Archaeology: Why the Past Matters*. Walnut Creek: Left Coast Press.

Lucas, G. 2001. *Critical Approaches to Fieldwork: Contemporary and Historical Archaeological Practice*. London: Routledge.

Mathers, C., T. Darvill, and B.J. Little. 2005. *Heritage of Value, Archaeology of Renown: Reshaping Archaeological Assessment and Significance*. Gainesville: University Press of Florida.

McCarthy, J.P., and J. Wickert. 2022. On the Horns of an Archaeological Dilemma: Balancing Site Confidentiality and Public Interpretation Imperatives at Delaware State Parks. In *Creating Participatory Dialogue in Archaeological and Cultural Heritage Interpretation: Multinational Perspectives*, ed. J.H. Jameson and S. Baugher, 67–83. New York: Springer.

McDavid, C. 2000. Archaeology as Cultural critique: Pragmatism and the Archaeology of a Southern States Plantation. In *Philosophy and Archaeological Practice: Perspectives for the 21st Century*, ed. C. Holtorf and H. Karlsson, 221–240. Göterborg: Göterborg Institutionen för arkeologi.

———. 2004. From 'Traditional' Archaeology to Public Archaeology to Community Action: The Levi Jordan Plantation Project. In *Public Archaeology as Applied Anthropology*, ed. P.A. Shackel and E.J. Chambers, 35–56. London: Routledge.

McGimsey, C.R. 1984. The value of Archaeology. In *Ethics and Values in Archaeology*, ed. E.L. Green, 171–174. New York: Free Press.

McGuire, R.H., and R. Paynter, eds. 1991. *The Archaeology of Inequality*. Oxford: Blackwell.

McManamon, F.P. 1991. The Many Publics for Archaeology. *American Antiquity* 56: 121–130.

Megaw, J.V.S., and D.D.A. Simpson. 1979. *Introduction to British Prehistory from the Arrival of Homo sapiens to the Claudian invasion*. Leicester: Leicester University Press.

Merriman, N. 1991. *Beyond the Glass Case: The Past, the Heritage and the Public in Britain*, Leicester Museum Studies Series. Leicester: Leicester University Press.

———. 2004. Introduction: Diversity and Dissonance in Public Archaeology. In *Public Archaeology*, ed. N. Merriman, 1–18. London: Routledge.

Milisauskas, S. 2002. *European Prehistory : A Survey*. New York NY/London: Kluwer Academic/ Plenum Publishers.

Mizoguchi, K. 2006. *Archaeology, Society and Identity in Modern Japan*, Cambridge Studies in Archaeology. Cambridge: Cambridge University Press.

Moshenska, G. 2008. Ethics and Ethical Critique in the Archaeology of Modern Conflict. *Norwegian Archaeological Review* 41 (2): 159–175.

———. 2017a. Alternative archaeologies. In *Key Concepts in Public Archaeology*, ed. G. Moshenska, 122–137. London: UCL Press.

———. 2017b. Introduction: Public Archaeology as Practice and Scholarship Where Archaeology Meets the World. In *Key Concepts in Public Archaeology*, ed. G. Moshenska, 1–13. London: UCL Press.

———., ed. 2017c. *Key Concepts in Public Archaeology*. London: UCL Press.

Moussouri, T. 2014. From 'Telling' to 'Consulting': A Perspective on Museums and Modes of Public Engagement. In *Public participation in archaeology*, ed. S. Thomas and J. Lea, 11–22. Woodbridge: Boydell Press.

Murray, T., and C. Evans, eds. 2008. *Histories of Archaeology: A Reader in the History of Archaeology*. Oxford: Oxford University Press.

Nativ, A. 2018. On the Object of Archaeology, with Comments by Lucas, G., Edgeworth, M., Witmore, C., and Harris, O. *Archaeological Dialogues* 25 (1): 1–47.

Ollich-Castanyer, I. (ed.) 2012. *Archaeology : New Approaches in Theory and Techniques*. InTech Open. Available online at https://www.intechopen.com/books/1999. Accessed 29 Mar 2023.

Patton, M. 2007. *Science, Politics and Business in the Work of Sir John Lubbock: A man of Universal Mind*. Aldershot: Ashgate.

Phillipson, D.W. 2005. *African Archaeology*. 3rd ed. Cambridge: Cambridge University Press.

Piggott, S. 1965. *Ancient Europe from the Beginnings of Agriculture to Classical Antiquity*. Edinburgh: Edinburgh University Press.

Potter, P.B. 1994. *Public Archaeology in Annapolis: A Critical Approach to History in Maryland's Ancient City*. Washingtom DC: Smithsonian Institution Press.

Prescott, C. 2013. Archaeology vs. Science; or Taking Knowledge-Based Communication Seriously? *Current Swedish Archaeology* 21: 39–44.

Quave, K.E., S.M. Fie, A.-S.Q.Q. Greiff, and D.A. Agnew. 2021. Centering the Margins: Knowledge Production in the Introductory Archaeology Course. *Advances in Archaeological Practice* 9 (2): 87–100.

Rathje, W.L., M. Shanks, and C. Whitmore, eds. 2013. *Archaeology in the Making: Conversations Through a Discipline*. London: Routledge.

Renfrew, C., and P. Bahn. 2016. *Archaeology : Theories, Methods and Practice*. 7th edn., Revised and Updated. London: Thames and Hudson.

Ribeiro, A. 2023. A Conversation with Koji Mizoguchi. On Globalization, Japanese Archaeology and Archaeological Theory Today. *Archaeological Dialogues* 1–21. Published online. https://doi.org/10.1017/S1380203823000016. Accessed 24 Mar 2023.

Richards, M., and K. Britton, eds. 2020. *Archaeological Science: An Introduction*. Cambridge: Cambridge University Press.

Roskams, S. 2001. *Excavation*, Cambridge Manuals in Archaeology. Cambridge: Cambridge University Press.

Sabloff, J.A. 2008. *Archaeology Matters: Action Archaeology in the Modern World*. Walnut Creek: Left Coast Press.

Sayce, A. H. 1976 [1884]. Preface. In Schliemann, H. 1976b [1884]. *Troja: Results of the Latest Researches and Discoveries on the Site of Homer's Troy, 1882*. Reprint edition. New York: Arno Press, v-xxx.

Scarre, G., and R. Coningham, eds. 2013. *Appropriating the Past: Philosophical Perspectives on the Practice of Archaeology*. Cambridge: Cambridge University Press.

Schadla-Hall, T. 2004. The Comforts of Unreason: The Importance and Relevance of Alternative Archaeology. In *Public Archaeology*, ed. N. Merriman, 255–271. London: Routledge.

Schliemann, H. 1976a [1881]. *Ilios: The City and Country of the Trojans. The Results of Researches and Discoveries on the Site of Troy and Throughout the Troad in the Years 1871, 72, 73, 78, 79*. Reprint edition. New York: Arno Press.

———. 1976b [1884]. *Troja: Results of the Latest Researches and Discoveries on the Site of Homer's Troy, 1882*. Reprint edition. New York: Arno press.

Seymour, D. J. (ed.) 2010. Special Issue: Black-and-White Issues About the Archaeological Grey Literature. *Archaeologies* 6(2).

Skeates, R., C. McDavid, and J. Carman, eds. 2012. *The Oxford Handbook of Public Archaeology*, Handbooks in Archaeology. Oxford: Oxford University Press.

Smith, C., and H. Burke. 2007. The Skin Game: Teaching to Redress Stereotypes of Indigenous People. In *Archaeology to Delight and Instruct: Active Learning in the University Classroom*, ed. H. Burke and C. Smith, 80–101. Walnut Creek: Left Coast Press.

Smith, L., and E. Waterton. 2009. *Heritage, Communities and Archaeology*, Duckworth Debates in Archaeology. London: Duckworth.

Stout, A. 2008. *Creating Prehistory: Druids, Ley Liners and Archaeologists in Pre-War Britain*. Oxford: Blackwell.

Swain, H. 2007. *An Introduction to Museum Archaeology*. Cambridge: Cambridge University Press.

Thomas, J. 2004. *Archaeology and Modernity*. London: Routledge.

Thomas, S. 2012. Archaeologists and Metal-Detector Users in England and Wales: Past, Present, and Future. In *The Oxford Handbook of Public Archaeology*, ed. R. Skeates, C. McDavid, and J. Carman, 60–81. Oxford: Oxford University Press.

Thomas, S., and J. Lea, eds. 2014. *Public Participation in Archaeology*. Woodbridge: Boydell Press.

Thomas, S., and P.G. Stone, eds. 2009. *Metal Detecting and Archaeology*. Woodbridge: Boydell Press.

Trigger, B.G. 2006. *A History of Archaeological Thought*. Cambridge: Cambridge University Press.

Uzzell, D.L. 1989. The Hot Interpretation of War and Conflict. In *Heritage Interpretation Vol. I: The Natural and Built Environment*, ed. D. Uzell, 33–47. London: Belhaven Press.

Venclová, N. 2007. Communication within Archaeology: Do We Understand Each Other? *European Journal of Archaeology* 10 (2–3): 207–222.

Virchow, R. 1976 [1881]. Preface. In Schliemann, H. *Ilios: The City and Country of the Trojans. The Results of Researches and Discoveries on the Site of Troy and Throughout the Troad in the Years 1871, 72, 73, 78, 79.* Reprint edition. New York: Arno Press, ix-xvi.

Walsh, K. 1992. *The Representation of the Past: Museums and Heritage in the Post-modern World*, The Heritage: Care-Preservation-Management. London: Routledge.

Watkins, J.E. 2006. Communicating Archaeology: Words to the Wise. *Journal of Social Archaeology* 6 (1): 100–118.

Chapter 5
Archaeology as Public Service

This chapter covers the area of work in which most archaeologists around the globe find employment, called variously—depending on choice, location or chronology—cultural resource management (CRM; Hatton and McManamon 2000), cultural heritage management (CHM; Messenger and Smith 2010a), archaeological resource management (ARM; McGimsey and Davis 1977), archaeological heritage management (AHM; Cleere 1984a, 1989a, b), development (or development-led) archaeology (Thomas 2013), *archéologie preventive* (Demoule 2012; Naffé et al. 2008) or sometimes simply public archaeology (PA; McGimsey 1972) although this latter term is more properly applied today to topics discussed in the second section of Chap. 4. Writers on public archaeologies (e.g. Skeates et al. 2012; Moshenska 2017) usually include work under these headings as part of what they cover because it is conducted in the public realm and notionally on behalf of a wider community, and because McGimsey's first published book on the topic was entitled *Public Archaeology* (McGimsey 1972) in the absence of any other label (although a mere five years later the term 'archaeological resources' would be applied: McGimsey and Davis 1977).

The range of terms has grown over the years since 1972, reflecting not only personal preferences but also different understandings of what exactly is being managed and even the problems of translation into different languages. The key differences until recently were between whether the material subject to management should be considered 'cultural' or 'archaeological', the key distinguishing feature being whether an extant living people can claim association with it (in which case it is 'cultural', as in the USA or Australasia) or not (in which case it is 'archaeological', as in Europe); and between whether it is a 'heritage' or a 'resource', which case the choice is more arbitrary because the argument for either tends to be couched in identical terms (see e.g. Messenger and Smith 2010b, 2; Hunter and Ralston 2006, vii–viii; McManamon and Hatton 2000). There are also problems of translation: a direct translation of 'cultural resource' into a Romance language (e.g. French, Italian or Spanish) makes no sense to those whose native languages they are: hence

J. Carman, *Archaeological Practice as Politics and Ethics*, https://doi.org/10.1007/978-3-031-69828-6_5

a preference for 'heritage' (which can be translated more meaningfully) over 'resource'. Everyone agrees, however, that whether culture or archaeology, heritage or resource, such material is necessarily subject to a need for its management.

The term '*archéologie preventive*' ('preventive archaeology') is a synonym for 'development(-led) archaeology' (Duong 2021) and largely reflects adoption across Europe of the international Convention advocating pre-emptive action in respect of remains likely to be damaged by building or other works (CoE 1992). It has been criticised on the grounds that the term suggests that it is archaeology that is being prevented rather than damage to archaeological material; and although the system was designed initially to identify remains that should be protected from damage by building and other works, it has in practice instead served to allow archaeologists access to sites for the purposes of investigation in the name of 'mitigating' the effects of work on buried remains. It was introduced to replace a previous reactive style of archaeology—which would intervene on the discovery of material during work and thus halt development projects—with a pro-active one which sought to identify the presence of any remains prior to work beginning; archaeologists would then conduct work to establish the nature and significance of such remains and once having done so depart to allow work to continue on the site. This approach is superseding the former reactive approach across the globe (see e.g. Naffé et al. 2008).

Whatever it is called, work of this kind is considered to be for a generalised public good, conforming to Merriman's (2004, 1–2) and my (Carman 2002, 97–112) distinction between archaeology 'in the public interest' or 'in the public realm' and engagement with actual people. The former involves working for, in or with governmental and other official bodies to comply with a set of regulations designed to meet the perceived needs of a wider community—most commonly national—in terms of preserving important aspects of its past as represented by physical remains. More broadly such aspects of the past are described as 'heritage' (see e.g. Carman 2002) and are represented by an 'authorised heritage discourse' (AHD; Smith 2006) which prescribes what is to be classed and treated as such a heritage. Archaeology as a public service conforms to the authorised discourse precisely because it is conducted as a generalised 'public good', subject to regulation, most usually by the state, and is conducted by or on behalf of public bodies, most often across the world branches of national government. It is this version of archaeology that is addressed in this chapter.

Governance by Legislation

Chapter 4 briefly mentioned the role of law in giving archaeologists control over their material and the power to limit any control exercised by others. Among others, Cleere (1989a, b, 10) goes further to argue that the archaeological heritage (his preferred term) is "governed by legislation", implying that no such thing exists without a legislative framework. Of course, archaeology (in the sense of the material with which we deal) has an existence independent of actual law: but as *heritage* or *resource* which is subject to systems of management it requires recognition by a regulatory framework. It is a truism that there is not only a body of law relating to its national patrimony including archaeological remains in every country in the world but also at the international level (see Carman 2015, 52–76). Table 5.1 shows only a sample of those from 1900CE to the mid-1980s as an illustration of how widely accepted is the idea of legislating for archaeology, but the practice has its origins in the medieval period of Europe (Carman 2015, 7) and its continuation since (Carman in press).

It is perhaps worth distinguishing the laws operative in nation-states from those operating at the international—regional, continental or global—level. The former have direct effect upon all those subject to them but only have force within the boundaries of the relevant state. The latter claim a wider transnational or universal applicability but do not have effect on individuals: the Conventions of organisations such as the Council of Europe, the Organisation of American States, etc. or of the United Nations or UNESCO are in fact agreements and treaties between groups of nation states and designed to ensure that the regulations and practices operative in those nation states meet the regional or global standards set by the Conventions. Such international regulation of fields such as archaeology is a relatively recent development, although treaties regulating other aspects of behaviour have a longer history, including peace treaties of various kinds and others designed to mitigate and control the use of weapons of war: cultural matters were specifically included in the agreements to end the series of wars in Europe from 1789 to 1815, and in the Hague Conventions on the conduct of international conflicts from the later nineteenth century onwards. The most notable of the latter is from 1954, specifically relating to matters of culture, and since that date a series of Conventions from international groups have sought to establish the principles upon which archaeology should be conducted (Table 5.1).

While elsewhere (especially in Carman 2015, 7–11, in press) I argue that these early legislative efforts—and indeed other practices contemporary with them—did not and were not intended to represent the management of archaeological material as we understand it today, nonetheless they represent practices that would be incorporated later into an emergent style for treating material of significance from the past. The literature of the field is always keen to emphasise that creating laws to control access to and use of archaeology is a practice sanctioned by age and long usage, so whatever these first efforts to claim control over ancient objects were intended to be, they are now held to be examples of what we do today, albeit incomplete and somewhat deficient. The difference with our practice is that these early legislative and regulatory programmes—whether from fourteenth or fifteenth century Italy or seventeenth century Scandinavia (Carman 2015, 6–11)—served the purposes of narrow interest, usually that of rulers, rather than any wider community; from the nineteenth century they served emergent nationalist or imperial interests (Carman 2015, 11–26); and only in the later twentieth century was there the development of a concern for the wider communities whose archaeology would be subject to management regimes under the heading of 'stewardship' (Carman 2015, 39–42).

Table 5.1 The global spread of modern archaeological legislation by country and date through the twentieth century to the 1980s

Region	1901–1920	1921–1940	1941–1960	1961–1980	1981+
Europe	Italy 1909 France 1909 Italy 1912 France 1913 Italy 1913 Great Britain 1913 Poland 1918 Austria 1918 France 1920	Italy 1922 Austria 1923 Malta 1925 Italy 1927 Poland 1928 Ireland 1930 Great Britain 1931 Belgium 1931 Great Britain 1932 Greece 1932 Poland 1933 Cyprus 1935 Italy 1939 Vatican City 1939	France 1941 Sweden 1942 Switzerland 1945 Great Britain 1947 Switzerland 1947 Poland 1948 Greece 1950 Great Britain 1953 Germany 1954 Ireland 1954 Germany 1955 Romania 1955 Czech Republic 1958 Austria 1958 Slovakia 1958	Slovenia 1961 Poland 1962 Finland 1963 Hungary 1963 Switzerland 1963 Cyprus 1964 Germany 1965 Yugoslavia 1965 Luxembourg 1966 Croatia 1967 Sweden 1967 France 1968 Bulgaria 1969 Iceland 1969 France 1970 Albania 1971 Sweden 1971 Belgium 1972 Italy 1972 Bulgaria 1973 Cyprus 1973 Romania 1974 Spain 1974 Hungary 1975 Poland 1976 Ukraine 1976 USSR 1976 Hungary 1977 Lichtenstein 1977 Serbia 1977 Bosnia 1978 Byelorussia 1978 Finland 1978 France 1979 Great Britain 1979 Norway 1979 Great Britain 1980	Gibraltar 1982 Great Britain 1983 Luxembourg 1983 Denmark 1984 Netherlands 1984 Portugal 1985 Spain 1985 Austria 1986 Denmark 1986 Finland 1986 Sweden 1986 Ireland 1987 Great Britain 1990

(continued)

Table 5.1 (continued)

Region	1901–1920	1921–1940	1941–1960	1961–1980	1981+
Western Asia and North Africa	Egypt 1912 Tunisia 1920	Turkey 1922 Iraq 1922 Iraq 1924 Iraq 1926 Palestine 1929 Iran 1930 Lebanon 1933 Lebanon 1934 Iraq 1936	Lebanon 1943 Syria 1947 Egypt 1951 Afghanistan 1958 Kuwait 1960	Syria 1963 Algeria 1967 Jordan 1968 Libya 1968 Bahrain 1970 UAE 1970 Yemen 1970 Saudi Arabia 1972 Yemen 1972 Turkey 1973 Iraq 1974 Iran 1975 Iraq 1975 Jordan 1976 Israel 1978 Morocco 1980 Qatar 1980	Egypt 1983 Israel 1983 Turkey 1983 Tunisia 1986
Central, South and East Asia	India 1906 Indonesia 1913	Cambodia 1925 Laos 1925 Vietnam 1925 China 1930 China 1931 Indonesia 1931 Thailand 1931 India 1932 Sri Lanka 1940 Mauritius 1944	Japan 1950 China 1950 Nepal 1956 Sri Lanka 1956 Burma 1957 India 1958 China 1960	Thailand 1961 Burma 1962 Nepal 1964 Brunei 1967 Pakistan 1968 Mongolia 1970 Nepal 1970 Singapore 1970 India 1972 Philippines 1973 Pakistan 1975 Bangladesh 1976 India 1976 Malaysia 1976 Pakistan 1979	China 1982 Laos 1982 Korea 1983 Taiwan 1984
Sub-Saharan Africa		Mali 1923 Benin 1937 Cameroon 1937 Chad 1937 Centrafrique 1937 Congo 1937 Comoros 1937 Djibouti 1937 Gabon 1937 Guinea 1937 Côte d'Ivoire 1937 Madagascar 1937 Mali 1937 Mauretania 1937	Benin 1944 Cameroon 1944 Chad 1944 Centrafrique 1944 Congo 1944 Comoros 1944 Djibouti 1944 Gabon 1944 Guinea 1944 Côte d'Ivoire 1944 Madagascar 1944 Mali 1944 Mauretania 1944 Morocco 1944 Niger 1944	Sierra Leone 1962 Cameroon 1963 Tanzania 1964 Malawi 1965 Ethiopia 1966 Lesotho 1967 Sierra Leone 1967 Uganda 1967 Ethiopia 1969 Ghana 1969 Lesotho 1969 South Africa 1969 Botswana 1970 Senegal 1971 Zaire 1971 Mauretania 1972	Kenya 1983 Burkino-Faso 1985 Mali 1985 Mali 1986 Côte d'Ivoire 1987

(continued)

Table 5.1 (continued)

Region	1901–1920	1921–1940	1941–1960	1961–1980	1981+
		Morocco 1937 Niger 1937 Réunion 1937 Senegal 1937 Togo 1937 Tunisia 1937 Haute Volta 1937 Burundi 1939 Rwanda 1939	Réunion 1944 Senegal 1944 Togo 1944 Tunisia 1944 Haute Volta 1944 Sierra Leone 1947 Zambia 1948 Sudan 1952 Benin 1956 Cameroon 1956 Chad 1956 Centrafrique 1956 Congo 1956 Comoros 1956 Djibouti 1956 Gabon 1956 Guinea 1956 Côte d'Ivoire 1956 Madagascar 1956 Mali 1956 Mauretania 1956 Morocco 1956 Niger 1956 Réunion 1956 Senegal 1956 Togo 1956 Tunisia 1956 Haute Volta 1956 Burundi 1956 Chad 1960	Swaziland 1972 Zimbabwe 1972 Ghana 1973 Madagascar 1973 Gambia 1974 Angola 1976 Nigeria 1979 Seychelles 1979 Tanzania 1979	
Caribbean, Central and North America	El Salvador 1903 USA 1906 Mexico 1912 Argentina 1913	Argentina 1921 Guatemala 1922 Costa Rica 1923 Canada 1927 Canada 1930 Mexico 1934 USA 1935 Colombia 1936 Costa Rica 1938 Haiti 1940	Haiti 1941 Venezuela 1945 Guatemala 1947 Canada 1953 Colombia 1959	USA 1966 Guatemala 1966 Honduras 1966 Colombia 1967 Dominican Republic 1968 Honduras 1968 USA 1969 Panama 1970 Belize 1971 Colombia 1971 Mexico 1972 Colombia 1973 Dominican Republic 1973 Mexico 1975 USA 1976 Cuba 1977 Cuba 1979 USA 1979	Canada 1975 Barbados 1981 Costa Rica 1982 Nicaragua 1982 Panama 1982 Honduras 1984 Cuba 1983 Jamaica 1985 Canada 1986 USA 1990

(continued)

Table 5.1 (continued)

Region	1901–1920	1921–1940	1941–1960	1961–1980	1981+
South America	Argentina 1911 Peru 1911	Argentina 1921 Peru 1929 Brazil 1937 Argentina 1940	Paraguay 1943 Ecuador 1945 Paraguay 1950 Suriname 1952 Peru 1958	Bolivia 1961 Brazil 1961 Chile 1961 Brazil 1965 Chile 1966 Chile 1969 Chile 1970 Uruguay 1971 Guyana 1972 Ecuador 1979	Peru 1985
Oceana and Australasia	New Zealand 1901 New Zealand 1908 Cook Islands 1915	Fiji 1929 Nauru 1935 Fiji 1940	Cook Islands 1950 New Zealand 1954 Samoa 1954 Australia 1955 Wallis & Futuna 1956 French Polynesia 1956 New Caledonia 1956 New Caledonia 1960	French Polynesia 1962 Papua New Guinea 1965 Vanuatu 1965 Niue 1966 Fiji 1968 Tonga 1969 New Zealand 1972 Australia 1975 New Zealand 1975 Micronesia 1979 Tuvalu 1979 New Zealand 1980 Australia 1980 Solomon Islands 1980	Mariana Isles 1982 Australia 1983 Guadalcanal 1983 Australia 1984 Australia 1986 Australia 1988
International[a]		Palestine Mandate[a] 1922	Hague 1954 European Culture 1954	European Heritage 1969 Paris 1970 World Heritage 1972 San Salvador 1976	

Source and for details of each: O'Keefe and Prott (1984, 1989), King (2008), and Messenger and Smith (2010a, b)

[a]Items marked with an asterisk denote bi- or multi-lateral agreements with limited extent; others in this set are international Conventions promoted by agencies such as UNESCO, the Council of Europe or Organisation of American States

Laws on archaeology perform a number of useful functions which assist but are not vital to their purpose. The functions, as revealed by a study of a number of European examples, are generally:

- To define the nature of the archaeology covered by the rules: especially the attributes and characteristics it must have to qualify for such status;
- To define how such material shall be identified and how its presence should be recorded in inventories;
- How such material shall be preserved and protected from damage or destruction, such as by designation as in some way special, or by controls on others' actions in relation to it;
- Definition of the type of conservation principle to be applied, including prescribing or proscribing reconstruction or restoration;
- Sanctions against breaches of the law, civil or criminal, and the means of ensuring compliance;
- Financial arrangements in relation to archaeological work;
- Integration of archaeological arrangements into other bodies of regulation, to ensure against contradiction and conflict;
- The powers and duties of officials, official agencies and those otherwise empowered to act in relation to archaeology;
- Provisions for formal or more general public education in archaeological matters (Pickard 2001, 4–10).

Not all laws everywhere will include matters on all these aspects, and there will be areas of the world where additional matters will be addressed in such laws, but these are the kinds of matters to be generally expected in such regulation. In all cases, the law in itself achieves nothing: instead it relies upon other agencies to ensure compliance and what the law states to be proper conduct in relation to archaeological material.

What laws do achieve simply by their existence is to classify the material they are concerned with as important. By regulating what can be considered as archaeological, by allocating its management and treatment to particular persons or agencies while denying it to others, and by asserting what kind of significance it has and appropriate uses to which it can be put, it separates all such material from the rest of the material in the universe and ensures that it is treated as something special. An inevitable side-effect, as noted especially by Smith (2004, 126–127), is to give authority to those who undertake the management of archaeological resources. Sometimes this is quite explicit: as noted in Chap. 1, the Valletta Convention (CoE 1992, Article 3ii) specifically requires that "excavations and other potentially destructive techniques are carried out only by qualified, specially authorised persons" which is widely interpreted to mean only those professionally qualified and accredited by relevant organisations; for those outside the Anglophone world it means most frequently accreditation by the State. Such requirements serve to establish the importance of archaeological material in two complementary ways: first, by focusing upon archaeology to the exclusion of other things and thereby separating it from all else; and secondly by requiring the accreditation of only certain kinds of person as allowed to treat with such material, thereby marking it as in need of specialist treatment (see also Carman 1996 for commentary on how English law gives value to archaeological remains).

A frequent justification for the special status granted archaeology under law is that it is a store of values. These may be indicated by a general term—such as archaeology being of 'significance' or 'importance'—or by the allocation of particular types of value such as 'scientific', 'historic', 'aesthetic', 'architectural', 'cultural', 'religious', etc. There may be levels of value: some may be classed as merely of '[some type of] interest' while others carry 'importance' or 'significance' which is presumably a higher grade of value. There may also be implied levels of value in the way certain things can be treated: so that anything of a certain age may be held to be of more importance than something younger; or items or places of 'cultural' significance (e.g. to a certain community) may be treated as if of greater importance than items or places of (mere) general historic interest. All such issues will vary from place to place and context to context, to suit local conditions and expectations. Nonetheless, the general idea that archaeological remains are to be considered as things of value is a universal principle underlying all efforts at regulation both at state and international level. The phrase 'are to be considered as things of value' is significant here: laws are not applied to archaeology because it *is* of value in advance of regulation, but rather it *acquires* that value because laws are applied to it. In other words, archaeological material—and the work of archaeologists on that material—is significant because of the way it is treated: it is not treated in a particular way because it has some inherent value.

The material covered by law can (*per* O'Keefe and Prott 1984, 184–187) be marked in one of three ways: enumeration; categorisation; and classification. *Enumeration* is a system of lists of the kinds of material to be covered: this is typical of US federal laws in this area (US Dept. of the Interior 1989–90) and has been to some extent adopted in the UK for the purpose of describing the kinds of objects that can be considered for the purposes of legal protection (Carman 1996, 120–124 and 187–192). The problem with this approach is that it leaves open the question of whether items not on the list but of a similar kind can be included: e.g. if the list specifies 'graves and burial sites', does this also cover above-ground disposal of the dead? *Categorisation* is a looser approach whereby a broad description of types of material is provided, into which a range of particular objects may fall. The problem of this approach is that too narrow a definition may exclude objects of concern, while too broad a definition may include too much material. By contrast with both, *classification* is not concerned with the form of the object, but with actions taken towards it: in such a system, only those objects officially recognised and designated as such by a responsible authority can be granted protection. While convenient and transparent, the system has the flaw of only recognising those objects that have been specifically designated, leaving others of similar nature to their fate. At the same time, it is worth noting that these different systems are by no means exclusive. It is possible to use them in combination, so that the list under an enumerative scheme may include categories, while a scheme of categorisation may also enumerate particular types of object, and a classificatory scheme may operate in respect of items enumerated or categorised.

These differences represent contrasting approaches to the archaeological resource as a phenomenon as well as the structure of law. Where only designated material is covered by law, the emphasis is placed upon the relevant authority and its decisions; where material is enumerated, anything included in the list is automatically covered, removing authority from agencies and placing it more generally; under schemes of categorisation, a measure of interpretation is required, placing some but not all focus upon agencies. An enumerative scheme assumes a solid understanding of the kinds of materials and places constituting the archaeological record: by its nature, anything not listed is excluded. A scheme of categorisation has a greater capacity for the inclusion of new types of material, especially if the categories are drawn not on the basis of physical form or attributes (e.g. state of ruination or age) but on value ascriptions (e.g. 'of architectural, archaeological etc. interest or importance' as discussed above). Paradoxically, the greatest flexibility may exist under a scheme of designation, so long as the capacity to designate is drawn widely: if it is limited by enumeration or categorisation, then it is significantly less able to include new types of material.

Despite the ubiquity of legislation as a foundational tool of archaeological resource management practice, very little of the literature of archaeology concerns the purpose of such laws; or, to put it another way, explains why we pass laws on this matter rather than tackling it in another way. McGimsey (1972), for instance, argues powerfully for legislation as a key component of a state preservation programme, but also argues against legislation alone since it would be an entirely "negative approach" (McGimsey 1972, 33 and 46) lacking the necessary support from the wider public. In so far as McGimsey does provide a reason for legislation, it must be as part of the requisite "administrative structure" (McGimsey 1972, 27) for such a programme, which includes its establishment as a legally-recognised authority with its own budget. O'Keefe and Prott (1984) go further: they argue that the dangers facing the archaeological resource are ever greater and that accordingly "some of them can only be controlled by governments" and therefore require legislation (O'Keefe and Prott 1984, 13). At the same time, they recognise the valuable role laws play in resolving key conflicts over material—especially issues of ownership and control—and the setting of policy aims; as well as the increasing requirements of national governments to comply with international treaties concerning the heritage (O'Keefe and Prott 1984, 14). None of these is, however, a reason for law *as such*: both McGimsey and O'Keefe & Prott also offer programmes of public education and the mustering of political support as alternatives (McGimsey 1972, 29–31; O'Keefe and Prott 1984, 145–150).

It has been argued that law sets "the parameters of acceptable management practice…., the scope of policy debate, and [thereby] influences the way in which debate is conducted" (Smith 2004, 125). Overall, "legislation provides governments and bureaucracies with terms, concepts and guidelines against which competing claims to material culture may be assessed" (Smith 2004, 126) and ultimately "provides the conceptual frameworks that must govern debates within" archaeological practice which "institutionalise and regulate the discipline [of archaeology] as a technology of government" (Smith 2004, 154). Similarly, Fourmile (1996) has reviewed the role of Australian legislation in denying the indigenous population any access to or control over their cultural heritage. These readings of the place of legislation in archaeology locate it at the service of requirements external to the discipline itself and closer to those of government. In other words, rather than law serving the needs of archaeology, archaeology is made to serve the needs of, and act as a tool, of government (on which see also Chap. 6).

Interestingly, however, it not just those who are critical (or indeed suspicious) of law who see it in this light. Breeze (1996)—writing on the definition given in Scotland to the British legislative category of 'national monument'—is clear that the purpose is "to ensure that all people have access to [Scotland's built] heritage [of all periods] and are able to enjoy it, regardless of their own origins and background" (Breeze 1996, 102). He also acknowledges that "preserving monuments… is not entirely an end in itself" and cites government reasoning behind it (Breeze 1996, 102). Accordingly, the idea of a 'national' archaeological resource based in law is seen here not as a limiting and exclusive concept but nevertheless one that remains at the service of government agendas. This same idea is reflected in Knudson's (1986) review of cultural resource management practice in the USA. As a result of success in "persuading the major policymakers… of the public significance of archaeological resources…. the implementation of such policies will not leave anyone… out of the process of public accountability for the treatment of those resources", and "this will be conducted within the context of multiple public objectives" (Knudson 1986, 399). The public referred to here is taken to be the Euro-American population of the USA, excluding its Indigenous population whose cultural works are under discussion. Accordingly, even though it is acknowledged that conservation of cultural remains is a globally-endorsed project, the target of conservation practice in the USA and what flows from it is directed at a particular audience, at least partly the result of "a lack of genetic continuity between the dominant political community in the United States and prehistoric Americans" (Knudson 1986, 396). Here, as in the UK, law drives the ARM process rather than providing support for it.

In most writing on ARM, a legislative basis for how we treat archaeological resources is taken for granted. The literature is therefore for the most part descriptive rather than critically discursive and does not ask why laws are in place in such profusion. One reason is simply historical: it is 'the way it is done' and has been since the medieval period and certainly since the establishment of nation states. Other reasons emerge from a closer reading, however, whether from an overtly critical or a more sympathetic perspective: as we have seen, laws serve, as it turns out, not the needs of archaeology, but rather the agencies—and in particular national governments—who promote them. This is not an issue of effectiveness, but has an impact on the way archaeology is done in different contexts since laws are to be read and applied differently in varying contexts. As I have argued elsewhere (Carman 1996, 17; 2002, 102–103; 2015, 57–59), to have a truly meaningful comparison between the practices of archaeological resource management in different territories it is necessary to take three factors into account:

- differences between legal and regulatory systems in different territories;
- differences in the nature of the material record of the past between one territory and another; and
- differences in the traditions and historical development of archaeology between one territory and the other.

The first of these covers such things as the basic assumptions relating to the interests to be served by law, the degree of appropriate state control held to be applicable in an area, the weight to be given to private property laws, or the expected powers and duties of state and other agencies. All of these will differ between one territory and another, or one legal system (e.g. Common or Roman) and another. In the UK or USA, for instance, the usual style is to provide for legal protection without taking material directly into state ownership, but in other territories all archaeological remains and other heritage objects are held to be the property of the state. In the UK, the USA and Australia this reflects the ideological authority of private property upheld by a system of Common Law; as against the authority of the state more typical of systems deriving from the European continent. Here, the difference lies in expectations of what is right and proper, and more fundamental social values. Where it is expected that heritage objects should belong to the state, the kind of ARM system operated in the UK or the USA—where archaeology can be carried out by private contractors or amateurs—makes no sense; in the UK or USA, the adoption of a system of generalised state ownership would be seen as an attack on private property. An attempt to assess the merits of one system against another therefore runs up against these fundamental differences in understanding of what laws can and should do, and to whom legal authority should be given (and from experience it is evident that this is often a problem in debates at archaeological conferences, where proponents of different systems so often talk 'past' each other because of a failure to grasp the fundamental differences of attitudes).

The second and third factors are linked. They concern the nature of the archaeological record and how it inevitably differs in different territories; and the understanding given to the purpose and focus of archaeological research which will differ in one country from another, so that very different research traditions may exist, leading to a differential emphasis on types of material. In the UK, for instance, the treatment of different types of material is very often the same regardless of physical form or age. Prehistoric structures in the countryside can be treated in exactly the same way as medieval ruins in a city; and ancient monuments (a legal category that in England now includes some material from the twentieth century) can be placed upon a schedule while standing buildings can be placed upon a list, both of which offer some form of legal protection. There are other territories, however, where differences in age make a substantial difference. Material from a preliterate past may be treated very differently from material emanating from historical times; or one period of history—or material representing a particular way of life—may be more highly valued than another, making one subject to legal control and protection while the other is abandoned to its fate. In the USA, for instance, buried remains of the Indigenous population are subject to forms of Federal legal control while the remains of (sometimes contemporary) colonising Europeans are excluded from this coverage. Such differences will make any direct comparison of UK and US laws rather meaningless, since they are grounded in very different historical circumstances, are driven by very different political and cultural imperatives, and concern significantly different categories of person. At root, therefore, UK and US legislation in this area do not concern the same types of material.

Any set of national laws will also need to be read in accordance with specific standards. These 'rules of construction' are quite precise and are often themselves enshrined in law, ensuring that any law of the particular state will be interpreted in the same way as any other and thus guarantee consistency in application. These rules do not, however, cross territorial and jurisdictional boundaries. A brief introduction to some of the key differences that can exist is set out by O'Keefe and Prott (1984, 150–151) and another by Summers and Taruffo (1991, 501) but for specific advice on how to read laws in particular jurisdictions, more precise legal guidance needs to be sought. In particular, there are gross differences between the manner of interpreting laws between systems of legal Codes and the principles of Common Law. All start from the premise that laws are written and composed of words: the question arises as to how to understand the meaning and intent behind certain words and phrases, what parts of the written text can be used as indications of how a law should be interpreted, and what extraneous material (if any) can be used to assist in this (and see Carman 2015, 59–63 for a more detailed discussion of this).

While the application of law is a universal practice, laws do not all work the same way everywhere and there are significant differences between one territory and another. Nonetheless, archaeology as *heritage* (or archaeological material as a *resource*) is, as Cleere (1989a, b, 10) claims, subject to governance by law and it is this governance that puts archaeology in the public realm as a public good and a public service. As a consequence, archaeologists employed as archaeologists inevitably work in the heritage sector, representing roles in that sector and adopting—at least in that work -- the values and attitudes of others in that sector.

Archaeologists as Workers on Heritage

Pickard's (2001, 4–10) list of the functions of laws quoted above also provides by extension a list of the kinds of work performed in the heritage sector by its practitioners. As workers in the heritage sector, these are also the functions of archaeologists, and can be reduced from the more detailed list to four essential processes that work in sequence, albeit with intermediary stages (see Carman 2015), which are:

- Recording and inventory
  - which excludes non-relevant material
- Evaluation and assessment
  - leading to decisions concerning the fate of material, and especially

- Preservation
  - of what is most valued (Cleere 1984b, 127), sometimes for
- Presentation and public access.

An alternative—proposed by Clark (2005, 321)—offers a slightly different set of stages in the form of a 'conservation strategy' for a place, defined as "a document that sets out the significance of a site, and how that significance will be retained in any future use, alteration, management, or repair":

- Identify stakeholders
- Understand the site
- Assess significance
- Identify how significance is vulnerable
- Set aims and objectives for retaining significance
- Implement the plan and use it in decision-making.

These sequences are both those of progressive decision-making and as a result, as well as being based upon the same idea, serve to replicate Lipe's (1984) model of a finite and gradually reducing 'cultural resource base' as material passes through the series of decision-points, especially those relating to value. It puts archaeologists in the position of those who make these determinations since we are deemed to be the experts on the basis of whose knowledge appropriate decisions can be based: we accordingly act as agents of the 'authorised heritage discourse' as defined by Smith (2006). This constitutes a very specific understanding of the role of archaeology as a field of endeavour, of the status of the material with which we engage, and indeed of what archaeology is ultimately *for*. It is clearly linked to the idea of governance by law, which also, as noted above, gives archaeologists a measure of authority and status.

The role of the archaeologist as heritage manager in assessing value should not be underestimated. Commentators such as Lipe (1984), Darvill (1995), Briuer and Mathers (1996), Carman (2015) and contributors to Mathers et al. (2005) and Smith et al. (2010)—among others—are all evidently and clearly aware of how archaeological practice in the public realm is a focus on the ascription of value and the consequences of doing so. Even where there is no explicit mention of value terms—such as 'importance' or 'significance'—the discussion of managing archaeological material turns on ideas of value and the assumption of its worthiness to be preserved, especially at public expense. We are therefore ultimately the arbiters of what can be considered important from the past and how it should be treated. This is the case whether we class ourselves as agents of the State or otherwise: 'activist' archaeologists who work on behalf of the disadvantaged (see Chap. 6) are as much a part of this process as any employee of a national or regional agency. As I have noted elsewhere, however (Carman 2015, 127), there is remarkably little practical advice given to archaeologists about how to make such decisions: there are lists of criteria to be applied galore and at both national and international scales (see e.g. Carman 2015, 107–127), but almost nothing on what archaeologists themselves should actually *do* to make the judgement. This places the burden entirely upon us as the relevant expert, and even though work by Startin (1993) in Britain established that in practice and even without firm guidelines, archaeologists working independently tended to make similar evaluations for purposes of preservation decisions, it nonetheless collapses to individual judgements, confirming the overall authority granted by official recognition of our status.

Evaluation is of course a stage in the overall decision-making process that comes after identification of what it is that will be subject to assessment. This stage—of recognition or inventory—must inevitably come before any evaluation can take place, but it too relies exclusively on the expertise of the archaeologist since we are the agents who determine what is our specific area of concern. It is (as I have outlined elsewhere: Carman 2015, 98) a process of categorisation—whereby features and objects are relabelled in archaeological terms, frequently in accordance with a standardised set of terms, so that they can be meaningfully compared with other features and objects falling within the same category. Quite often the categorisation scheme matches one established by governing legislation, but this is not inevitable: as scientists (see Chap. 1) we have the capacity to develop our own schemes for our own purposes, especially in the application of digital technology (see e.g. Chadburn 1988; Beebe 2017). The identification and recording of sites, features and materials is of course one function of fieldwork (see Chap. 2) and derives directly from that, and it matters little (or not at all) whether that fieldwork has been carried out in the first instance for the purposes of resource management, as pure research, or for training of new recruits to the discipline. What matters is that there is a record of what has been identified that can then be used—if desired or required—for the purposes of evaluation.

It can be argued that once identification and evaluation have been carried out, the resource passes out of the remit of the archaeologist and becomes the object of attention by others whose task is more directly concerned with its future treatment: call these 'conservationists' or 'heritage managers' if you like. But as others have noted—especially Waterton and Smith (2010)—archaeologists are frequently those specifically employed in the agencies responsible for the preservation of the resource and their task becomes ensuring the survival of that resource into the future. The method to be applied is usually one of monitoring what happens to the object of preservation and putting in place appropriate limitations to damage or destruction: archaeologists rarely (if ever) stand over a site for twenty-four hours a day to protect it; instead they will seek to apply laws and regulations that punish damage and either they or others will be charged with site visits to check on its condition. This too is a form of fieldwork (Chap. 2)—closer to survey than excavation—but limited in scope and designed exclusively for the purposes of resource management. It can however become in effect a return to the 'identification' and 'inventory' stage if the condition of the site has altered such that it requires further evaluation. And so goes the process, once again echoing Lipe's (1984) ongoing and ever-reducing spiral.

Not every preserved site, feature or structure will become available for access by a visiting or viewing public and it can be argued that presentation is not a specifically archaeological function: again it is a role for those designated as 'heritage managers' or 'curators'. Nonetheless, it is the archaeological interpretation that will be presented to the public, and the archaeological literature abounds with discussions of making material available to the publics for whom we work (see e.g. Potter 1994; Molyneaux and Stone 1995; Copeland 2004; Parker Pearson and Pryor 2006; Matthews 2011; Aston 2012; Gustaffson and Karlsson 2012; Moussouri 2014; Grima 2017; etc.). We take the role of public advocate—both for our material and our work—very seriously and as a result accept the task of presenting both to a wider audience (see also Chap. 3). Nonetheless, as the heritage field increasingly becomes a specialist activity of those specifically trained and educated to undertake it, engagement with the public also becomes a specialism in its own right. Many institutions responsible for public engagement—national agencies, museums, etc.—employ specialist staff, very often trained as educators, to undertake the role (Cracknell and Corbishley 1986). Where others become involved, they find themselves increasingly involved in this aspect and more distanced from investigation of the past itself (e.g. McDavid 2000; Pryor 1989). The consequence is that—despite the best intentions of those involved—public outreach becomes something apart from the mainstream of archaeological activity. As Hills and Richards (2006) and Parker Pearson and Pryor (2006) demonstrate for Britain, the means of disseminating ideas and understanding of the subject are increasingly removed from the hands of those who engage directly with preserving the material past.

Gould (2018, 31–34) emphasises the engagement of archaeology as heritage in attempts to promote economic development. He explains that "development involves the integration of numerous complex physical, financial, and social systems into a working, self-sustaining process [and that] the development of historic places requires enormous capital and considerable technical expertise in conservation, marketing, planning, and operational management" (Gould 2018, 33), but that these are rarely available in local communities. The burden necessarily then falls upon experts to deliver and the voices—and indeed actual needs—of the community supposedly being served are lost, while local and national élites take the benefits because they are able to mobilise the resources needed to conduct the development project (Gould 2018, 33). And yet "without those experts… development cannot happen at all" (Gould 2018, 33). This necessarily puts the archaeologist at the service not of a public composed of living human beings but rather the generalised 'public good' identified earlier (and by Merriman 2004, 1–2; Carman 2002, 97–112) and which operationalises Smith's (2006) 'authorised heritage discourse'.

It will be evident that the involvement of archaeologists in these officially-driven heritage processes necessarily engages them in activity deeply informed by and supportive of established and normative political structures (and see Chap. 6). Such processes work to endorse the existence of current political structures such as the nation-state (as indicated in previous chapters), and indeed particular forms of government, by a focus on particular aspects of the past—and the material culture representing them—at the expense of others. Hamilakis (2007) is critical of the failure of those archaeologists who do not realise how their work is inevitably a political exercise, by not acknowledging "the power dynamics, asymmetries and inequalities… in the broader world [and requiring] the adoption of an active stance in today's clashes and battlegrounds" (Hamilakis 2007, 32). Similarly, Henderson (2007, 250) emphasises the role of heritage (including archaeology) "to convey a unifying sense of national identity and history and affirm control mechanisms.... and where [there is] a compulsion to trumpet the superiority of [government] doctrine and the benefits of life" under a particular regime. For Hamilakis and Duke (2007) there is a direct connection between the politics of capitalism and ethics which also relates to the professionalisation of archaeology, especially at the service of this amorphous and political public good.

As servants of the public good and the AHD, employed to provide specialist expertise by those concerned to meet the needs of government and of economic development, archaeologists are given status and authority. This is, however, only of a certain kind, which is constrained by the kind of work and the nature of the employer because it is to the latter that duty is owed: if to a State agency, the primary duty is that of meeting the needs of policy; if to a commercial developer, the primary duty is ultimately that of assisting them in making profit. Archaeologists are also expected to exercise the authority and the expertise they possess in a way that meets the ethical requirements of the field as set by archaeological associations and professional organisations, balancing among others the various interests listed in Table 1.1 and in the EAA (2022) statement of ethics with that primary duty to employer, regardless of their own individual political and moral beliefs and imperatives. For critics of such a 'commercialised' heritage (Hamilakis and Duke 2007; McGuire 2008; and see also Tarlow 2001), ethics is a broader concept that goes beyond doing the work for which one is paid in a manner that meets the needs of an employer, to a concern for the broader implications of the entire project in which one is engaged. As discussed in Chap. 1, these extend beyond the immediate audience for results of work, and also involve engagement with the needs of others who may not be immediate stakeholders: archaeologists who are employed in the service of dictatorial regimes or unscrupulous private entrepreneurs can be accused of helping maintain them in their position. It is a concerning proposition when so many of our colleagues across the world may be in exactly that position and it places a further ethical burden on those of us who are luckier in our situation.

Official Persons and Agencies

Employment of archaeologists in the public realm serves to provide the basis for the range of different functions archaeologists can perform as those responsible for various aspects of managing archaeological material as part of the cultural heritage. Depending where in the world you are located, they may take different forms and your employer may be one of a number of types of organisation. Most commonly outside the Anglophone world it will be the State or a state agency of some kind since the State will be the arbiter of what constitutes 'public service'. But—agent of the AHD or not—other possibilities also exist. Schofield et al. (2011, 100–105) offer three roles for the professional archaeologist in Britain, each of which will have its equivalent elsewhere even though the system of archaeology may not be quite the same. They are designated the 'three Cs' because of a shared initial letter.

- *Curators* are generally employed in the public sector by relevant agencies; elsewhere they may be direct employees of the State. Their role is to ensure proper management of the archaeological resource, to ensure preservation of significant remains and to work with others to mitigate any potential damage. In some places their role may be combined with that of others but in the purely curatorial role it is about protection of archaeology from damage.
- *Contractors* are those employed to carry out work—especially survey and investigation—on site. In many Anglophone countries, and increasingly across the world, they operate in independent commercial organisations who are paid to conduct this work. Elsewhere they may combine the role with that of curator if employed by the State or its agencies but the primary role is that of investigator.
- *Consultants* are those employed to advise others on matters relating to the archaeological resource and to liaise with others concerned. In Britain, the consultant is most likely to be employed by a developer to advise on what may be required to be able to complete a building project, and the consultant will therefore liaise with the relevant curator and act as agent in hiring a relevant contractor to conduct any necessary work. In other territories, the consultant may also be the contractor and sometimes the curator too: but as consultant the task is to work to balance ensuring proper treatment of the archaeological resource with other imperatives, whether of policy or commerce.

There is also a fourth 'C' (Schofield et al. 2011, 105–106) but this is the role of client, the non-archaeologist who will employ the contractor and consultant to minimise interference in their plans from the curator and for whom contractor and consultant will work: depending on the territory and specifics of the project, the client might be a commercial developer or the State or one of its agencies. For academic research projects, there will be no client as such: here, all these four roles will fall on the project director. However, as so-called preventive archaeology (discussed above) spreads across the globe, each of these roles and their separation one from another will become more common everywhere.

The specific title given to an archaeologist working in the public sector will vary depending on the territory. It will depend very much on the governmental system in place: a heavily centralised governmental organisation is likely to employ individuals directly, locating them regionally, each with a common title; a federal state will have separate arrangements in each state or province and titles may vary; a unitary state may have a single national body responsible for overall policy but also delegate responsibility to local governments, each of which may operate slightly differently. It will also depend on the institutions out of which archaeology is conducted: some states will have a dedicated archaeological or heritage service to which the only right and authority to conduct archaeological work will be given; others will depend on archaeologists based in museums or academic institutions, who will also be State employees; others will have a privatised commercial system in place, and where archaeologists operate or are employed by specialist firms providing archaeological services. In some states, the possibilities will be mixed, with some or all of these working together or in competition. All of course represent professional archaeologists: there is little opportunity for the amateur or avocational in official public activity.

Chapter 2 introduced early examples of individuals specifically employed to conduct work we can call archaeological—e.g. Raphael's investigations for a sixteenth century Pope, Henry VIII's 'King's antiquary' in sixteenth century England, Bure and Worm in seventeenth century Scandinavia (Carman 2015, 8–9), and the Parkers of Heytesbury in late eighteenth and early nineteenth century Britain (Everill 2012). The later development of specific positions in government service across Europe (Chamberlin 1979) and the establishment of organisations such as the Archaeological Survey of India in the mid-nineteenth century (Carman 2015, 16), among others, further established the idea of a profession of archaeology, although the recognition of archaeology as a profession is a specifically late twentieth century phenomenon (to be discussed in Chap. 7). Here, the significant point is that these functions operate as part of an official approach to the management of archaeological resources, and that there are organisations across the world dedicated to that purpose who act as employers for archaeologists. The majority of employing organisations operate at the national level because it is primarily at the national and more local level that archaeology as a function of heritage management operates (see e.g. Cleere 1984a; Messenger and Smith 2010a, b; Hunter and Ralston 2006; Sebastian and Lipe 2009; Carman 2015), reaffirming the close connection we have seen in other chapters between archaeology and the nation-state.

There is little difference in the roles required in a state-run archaeological service and a privatised one. The three Cs identified above—curator, contractor and consultant—will have their correlates wherever such work is carried out. As indicated, in a state-run system they may be combined into fewer individuals, so that the person responsible for ensuring proper treatment of archaeological resources (= the curator) may also be the one who carries out—or more likely directs—any fieldwork (= the contractor) and may also be the person who advises other stakeholders on what they may or may not do (= the consultant). There will be a difference of emphasis from a privatised system: less liaison between actors and more issuing of instructions; more work as curator and contractor than as consultant; and a significant wielding of authority over non-state persons and institutions. It may be more reactive if a system of preventive archaeology is not yet in place: involvement of archaeologists may not take place except and until archaeological remains are discovered in work on site and in turn relies upon the discoverers reporting it to the relevant authority. There may be more of a focus on monitoring the condition of known sites and monuments, putting one aspect of the curatorial role above others, rather than examining development projects for likely effects on any remains or conducting fieldwork and excavation. All such work will be conducted for and on behalf of the state, generally to the exclusion of other interest, since it will be the state that takes sole responsibility for the archaeological resource.

In a privatised—or partly privatised—system the three roles are more likely to be distributed among different persons. The curatorial role will be conducted by those employed at the service of a government, or branch thereof, or an agency working on behalf of the government. It may involve not only the monitoring of the condition of known sites and monuments but also and especially a concern with proposed building and other works—agriculture or quarrying—that may affect any buried remains. It may require the exercise of a level of authority to prevent or modify such plans so as to mitigate their effects and some liaison with the developer to ensure compliance. It may also involve the employment of a contractor to carry out any necessary investigation of remains and to establish their nature, condition and significance: although in some territories the contractor is employed by the developer, tenders for the work having been adjudicated by the curator to ensure work will be conducted to the appropriate standards. The curator may also monitor the work of the contractor, and reports on the work—most commonly examples of 'grey literature' (as discussed in Chap. 4)—will be submitted to the curator. The consultant, if employed, will act for the developer as an independent commentator on the work of the curator and contractor. They may liaise with the curator over what work—if any—is required, and how it is best conducted, and may be involved in the selection of contractor, with whom they will also liaise in a supervisory capacity for the developer. The three roles will together collaborate in the service of the archaeological resource, each of them having to compromise on some issues, and it will depend on attitudes towards private property, state interference in private business and how important is the past and its material remains as to how much authority and control each can wield.

The system of preventive archaeology spreading across the globe can be combined with either a state-run or privatised system, although it tends to encourage privatisation and was created to serve such arrangements in the first instance (Demoule 2012). The aim as indicated earlier, is to identify remains that may be affected—and especially damaged—by work on the land in advance of the work commencing, allowing mitigation to be put into place. In practice, as also noted earlier, it has become a system allowing archaeologists access to remains—and therefore work—since it is assumed that the significance, or indeed form, of remains cannot be known without investigation. In a state-run system, the initial identification of the possible existence of remains, its investigation, and decisions about its future will all fall on the same people, all of them employed by the state or its agency. In a privatised system, as suggested above, the roles will be divided among different persons who act for and serve different organisations: the state or a state agency and the developer. Nonetheless, the end result is intended to be the same: the preservation of significant remains by ensuring its identification before destruction, such preservation to be either *in situ* (i.e. physical preservation) or 'by record' in the form of a report and archive (Chap. 3).

Differences in system—exclusive state control or privatisation, pro-active 'preventive archaeology' or a more reactive one—says nothing about effectiveness: any system will work so long as that is the one that meets the understandings and requirements of the local context. Where it is generally assumed that the state is the only safe arbiter of archaeological decision-making, delegation of authority to non-state actors will be counter-productive; where exclusive state authority would be treated as an assault on professional privilege and private property, any attempt to assert such control will be resisted. This is not to deny the possibility of change: preventive developer-led archaeology is being adopted increasingly across the globe, leading also to the establishment of non-state organisations to undertake archaeological work: the effectiveness of such new arrangements depends upon other factors, not least the numbers of archaeologists employed to undertake work.

While in the global North, individual countries and groups of countries such as the European Union may be able to marshal archaeologists in their thousands, in the global South they may be far fewer and some of the world's poorer territories will have none to call upon because the educated will be required for to meet more urgent functions than preservation of a distant past. In such a case, it would make very little difference whether the system were state-driven, privatised, reactive or pro-active: there will be no-one to undertake the work unless they can be brought in from outside. It follows that in assessing the effectiveness of archaeology anywhere, there are a range of factors to be considered, not the least of which is the education system in place to educate and train archaeologists together with the political will to devote resources to such education.

McGimsey's (1972, 5) first full-length treatment of the topic of this chapter states at the outset that "there is no such thing as a 'private' archaeology" and his book is a justification for and description of a public system of cultural resource management created for the needs of a particular US state. For him, therefore, all archaeology is for the public good, and that public good is mobilised through the establishment of laws and structures designed to give archaeologists control over the material with which they are concerned. This was an entirely novel idea—it underpinned all official state-driven efforts to manage pasts by gaining control over its material remains, from medieval Popes in Italy through to modern nation-states across the globe. The official agencies and appointments in which and through which archaeologists can manage the resource is merely the means by which this public service ideal is achieved. As we shall see in the next chapter, there are other senses in which the idea of archaeology as a 'public good' is expressed (e.g. by Green et al. 2021). However, the general public good—one not intended to be limited to specific audiences or communities—is the one aimed to be served by official activity in archaeology at the local, national and international level. That it serves other purposes too—such as the maintenance of certain political forms, particular structures of governance, and the authority of classes of person including the professional archaeologists—is a by-product (happy or unhappy, depending on point of view) which we need to recognise. While accepting our role as servants of the public good, it also falls to us to identify that the means we use also have other effects, and to make a choice as ethical practitioners of whether we wish to perpetuate or challenge them. That is how archaeology is inevitably also a political process (an issue to be addressed in Chap. 6).

References

Aston, M. 2012. Publicizing Archaeology in Britain in the Late Twentieth Century: A Personal View. In *The Oxford Handbook of Public Archaeology*, ed. R. Skeates, C. McDavid, and J. Carman, 373–394. Oxford: Oxford university Press.

Beebe, C. 2017. Standard Descriptive Vocabulary and Archaeology Digital Data Collection. *Advances in Archaeological Practice* 5 (3): 250–264.

Breeze, D. 1996. Archaeological Nationalism as Defined by Law in Britain. In *Nationalism and Archaeology*, Scottish Archaeological Forum, ed. J.A. Atkinson, I. Banks, and J. O'Sullivan, 95–103. Glagow: Cruithne Press.

Briuer, F.L., and W. Mathers. 1996. *Trends and Patterns in Cultural Resource Significance: An Historical Perspective and Annotated Bibliography*. Alexandria: US Army Corps of Engineers.

Carman, J. 1996. *Valuing Ancient Things: Archaeology and Law*. Leicester: Leicester University Press.

———. 2002. *Archaeology and Heritage: An Introduction*. London: Continuum Press.

———. 2015. *Archaeological Resource Management: An International Perspective*. Cambridge: Cambridge University Press.

Carman, J. in press. 'Governed by Legislation': What Laws Do to Archaeology. In Diaz-Andreu, M. & Coltofean, L. (eds) *Oxford Handbook on the History of Archaeology*. Oxford: Oxford University Press.

Chadburn, A. 1988. A Review of Approaches to Controlling Archaeological Vocabulary for Data Retrieval. In *Computer and Quantitative Methods in Archaeology 1988. CAA88*, BAR International Series 446 (ii), ed. S.P.Q. Rahtz, 389–398. Oxford: British Archaeological Reports.

Chamberlin, E.R. 1979. *Preserving the Past*. London: Methuen & Co.

Clark, K. 2005. The Bigger Picture: Archaeology and Values in Long Term Cultural Resource Management. In *Heritage of Value, Archaeology of Renown: Reshaping Archaeological Assessment and Significance*, ed. C. Mathers, T. Darvill, and B.J. Little, 317–330. Gainesville: University Press of Florida.

Cleere, H.F., ed. 1984a. *Approaches to the Archaeological Heritage*. Cambridge: Cambridge University Press.

———. 1984b. World Cultural Resource Management: Problems and Perspectives. In *Approaches to the Archaeological Heritage*, ed. H.F. Cleere, 125–131. Cambridge: Cambridge University Press.

———., ed. 1989a. *Archaeological Heritage Management in the Modern World*, One World Archaeology 9. London: Unwin Hyman.

———. 1989b. Introduction: The Rationale of Archaeological Heritage Management. In *Archaeological Heritage Management in the Modern World*, One World Archaeology 9, ed. H.F. Cleere, 1–19. London: Unwin Hyman.

Copeland, T. 2004. Presenting Archaeology to the Public: Constructing Insights On-Site. In *Public Archaeology*, ed. N. Merriman, 132–144. London: Routledge.

Council of Europe. 1992. *European Convention on the Protection of the Archaeological Heritage (Revised) (ETS No. 143)*. Available at: https://rm.coe.int/168007bd25. Accessed 17 Apr 2023.

Cracknell, S., and M. Corbishley, eds. 1986. *Presenting Archaeology to Young People*, CBA Research Report 64. London: Council for British Archaeology.

Darvill, T. 1995. Value Systems in Archaeology. In *Managing Archaeology*, ed. M.A. Cooper, A. Firth, J. Carman, and D. Wheatley, 40–50. London: Routledge.

Demoule, J.-P. 2012. Rescue Archaeology: A European View. *Annual Review of Anthropology* 41: 611–626.

Duong, M. 2021. Development-Led or 'Preventive' Archaeology in the Grand Duchy of Luxembourg, *Internet Archaeology* 57. https://doi.org/10.11141/ia.57.6. Accessed 13 Apr 2023.

European Association of Archaeologists (EAA). 2022. *EAA Annual Meeting Business Meeting Reports 2022*. Prague: European Association of Archaeologists. Available at: e-a-a.org. Accessed 23 Nov 2022.

Everill, P. 2012. *The Invisible Diggers: A Study of British Commercial Archaeology*. 2nd revised edition. Oxford: Oxbow.

Fourmile, H. 1996. The Law of the Land: Whose Land? In *Issues in Management Archaeology*, ed. A. Clark and L. Smith, 45–50. St Lucia: Tempus Publications, University of Queensland.

Gould, P.G. 2018. *Empowering Communities Through Archaeology and Heritage: The Role of Local Governance in Economic Development*, Debates in Archaeology. London: Bloomsbury.

Green, S., C. Green, and J. Schulderein. 2021. Archaeology as a Public Good. *Archaeologies: Journal of the World Archaeological Congress* 17 (1): 1–18.
Grima, R. 2017. Presenting Archaeological Sites to the Public. In *Key Concepts in Public Archaeology*, ed. G. Moshenska, 73–92. London: UCL Press.
Gustaffson, A., and H. Karlsson. 2012. 'Changing of the Guards': The Ethics of Public Interpretation of Cultural Heritage Sites. In *The Oxford Handbook of Public Archaeology*, ed. R. Skeates, C. McDavid, and J. Carman, 478–495. Oxford: Oxford University Press.
Hamilakis, Y. 2007. From Ethics to Politics. In *Archaeology and Capitalism: From Ethics to Politics*, ed. Y. Hamilakis and P. Duke, 15–40. London: Routledge.
Hamilakis, Y., and P. Duke, eds. 2007. *Archaeology and Capitalism: From Ethics to Politics*. London: Routledge.
Hatton, A., and F.P. McManamon. 2000. *Cultural Resource Management in Contemporary Society: Perspectives on Managing and Presenting the Past*, One World Archaeology 33. London: Unwin Hyman.
Henderson, J.C. 2007. Communism, Heritage and Tourism in East Asia. *International Journal of Heritage Studies* 13 (3): 240–254.
Hills, C., and J.D. Richards. 2006. The Dissemination of Information. In *Archaeological Resource Management in the UK: An Introduction*, 2nd edition (revised), ed. J. Hunter and I. Ralston, 304–315. Stroud: Sutton.
Hunter, J., and I. Ralston, eds. 2006. *Archaeological Resource Management in the UK: An Introduction*, 2nd edition (revised). Stroud: Sutton.
King, T.F. 2008. *Cultural Resource Laws & Practice*. New York: Altamira Press.
Knudson, R. 1986. Contemporary Cultural Resource Management. In *American Archaeology Past and Future: A Celebration of the Society for American Archaeology 1935–1985*, ed. D.J. Meltzer, D.D. Fowler, and J.A. Sabloff, 395–414. Washington, DC: Smithsonian Institution.
Lipe, W. 1984. Value and Meaning in Cultural Resources. In *Approaches to the Archaeological Heritage*, ed. H.F. Cleere, 1–11. Cambridge: Cambridge University Press.
Mathers, C., T. Darvill, and B.J. Little, eds. 2005. *Heritage of Value, Archaeology of Renown: Reshaping Archaeological Assessment and Significance*. Gainesville: University Press of Florida.
Matthews, C.N. 2011. Special Issue: Dynamics of Inclusion in Public Archaeology. *Archaeologies* 7 (3).
McDavid, C. 2000. Archaeology as Cultural Critique: Pragmatism and the Archaeology of a Southern States Plantation. In *Philosophy and Archaeological Practice: Perspectives for the 21st Century*, ed. C. Holtorf and H. Karlsson, 221–240. Göterborg: Göterborg Institutionen för arkeologi.
McGimsey, C.R. 1972. *Public Archaeology*. New York: Seminar Books.
McGimsey, C.R., and H.A. Davis. 1977. *The Management of Archaeological Resources: The Airlie House Report*. Washington, DC: Society for American Archaeology.
McGuire, R.H. 2008. *Archaeology and Political Action*. Berkeley: University of California Press.
McManamon, F.P., and A. Hatton. 2000. Introduction: Considering Cultural Resource Management in Modern Society. In *Cultural Resource Management in Contemporary Society: Perspectives on Managing and Presenting the Past*, One World Archaeology 33, ed. A. Hatton and F.P. McManamon, 26–62. London: Unwin Hyman.
Merriman, N. 2004. Introduction: Diversity and Dissonance in Public Archaeology. In *Public Archaeology*, ed. N. Merriman, 1–18. London: Routledge.
Messenger, P.M., and G.S. Smith, eds. 2010a. *Cultural Heritage Management: A Global Perspective*. Gainesville: University Press of Florida.
———. 2010b. Introduction. In *Cultural Heritage Management: A Global Perspective*, ed. P.M. Messenger and G.S. Smith, 1–7. Gainesville: University Press of Florida.
Molyneaux, B.L., and P.G. Stone. 1995. *The Presented Past: Heritage, Museums and Education*, One World Archaeology 25. London: Unwin Hyman.
Moshenska, G., ed. 2017. *Key Concepts in Public Archaeology*. London: UCL Press.

Moussouri, T. 2014. From 'Telling' to 'Consulting': A Perspective on Museums and Modes of Public Engagement. In *Public Participation in Archaeology*, ed. S. Thomas and J. Lea, 11–22. Woodbridge: Boydell Press.

Naffé, B.O.M., R. Lanfranchi, and N. Schlanger, eds. 2008. *L'archaéologie preventive en Afrique: enjeux et perspectives*. Saint-Maur-des-Fossées: Éditions Sepia.

O'Keefe, P.J., and L.V. Prott. 1984. *Law and the Cultural Heritage: Volume 1: Discovery and Excavation*. Abingdon: Professional Books.

———. 1989. *Law and the Cultural Heritage: Volume 3: Movement*. London: Butterrworths.

Parker Pearson, M., and F. Pryor. 2006. Visitors and Viewers Welcome? In *Archaeological Resource Management in the UK: An Introduction*, ed. J. In Hunter and I. Ralston, 2nd ed., 316–327. Stroud: Allan Sutton and IFA.

Pickard, R. 2001. Introduction. In *Policy and Law in Heritage Conservation*, ed. R. Pickard, 1–11. London/New York, Spon.

Potter, P.B. 1994. *Public Archaeology in Annapolis: A Critical Approach to History in Maryland's Ancient City*. Washington, DC: Smithsonian Institution Press.

Pryor, F. 1989. Look What We've Found – A Case Study in Public Archaeology. *Antiquity* 63: 51–61.

Schofield, J., J. Carman, and P. Belford. 2011. *Archaeological Practice in Great Britain: A Heritage Handbook*. New York: Springer.

Sebastian, L., and W.D. Lipe, eds. 2009. *Archaeology and Cultural Resource Management: Visions for the Future. School for Advanced Research Seminar*. Santa Fe: NM: SAR Press.

Skeates, R., C. McDavid, and J. Carman, eds. 2012. *The Oxford Handbook of Public Archaeology*, Handbooks in Archaeology. Oxford: Oxford University Press.

Smith, L. 2004. *Archaeological Theory and the Politics of Cultural Heritage*. London: Routledge.

———. 2006. *Uses of Heritage*. London: Routledge.

Smith, G., P.M. Messenger, and H.A. Soderland, eds. 2010. *Heritage Values in Contemporary Society*. Walnut Creek: Left Coast Press.

Startin, B. 1993. Assessment of Field Remains. In *Archaeological Resource Management in the UK: An Introduction*, ed. J. Hunter and I. Ralston, 1st ed., 184–196. Stroud: Allan Sutton and IFA.

Summers, R.S., and M. Taruffo. 1991. Interpretation and Comparative Analysis. In *Interpreting Statutes: A Comparative Study*, ed. D.N. MacCormick and R.S. Summers, 461–510. Aldershot: Dartmouth Publishing.

Tarlow, S. 2001. Decoding Ethics. *Public Archaeology* 1: 245–251.

Thomas, R.M. 2013. Making the Most of Development-led Archaeology. *The Archaeologist* 89: 12–16.

U.S. Department of the Interior. 1989–90. *Federal Historic Preservation Laws*. Edited by Blumenthal, S.K. Washington: US Department of the Interior.

Waterton, E., and L. Smith, eds. 2010. *Taking Archaeology Out of Heritage*. Cambridge: Cambridge Scholars Press.

Chapter 6
Archaeology as Politics

The last chapter could not avoid an emphasis on the essentially political nature of archaeology and this chapter continues to consider that aspect of the field. As a topic it reflects the other underlying themes of the book—especially the importance of archaeology as a set of practices in the world; and issues of ethics (interpreted more broadly than as codes of practice) as they affect the conduct of archaeological work. Archaeology as a form of political action—as indicated in Chap. 5—can serve either to support normative and established political forms or to challenge them and create new ones. That political action can be of a passive nature—merely doing archaeology that has as a consequence a political outcome which is not intended by the archaeologist—or an active one, in which the archaeological work itself is intended one way or another to have a political effect. Any form of archaeology as political action—passive or active—can serve to support established structures of power and authority or to challenge them: there is no automatic political result of archaeology as archaeology; it depends on the archaeological activity engaged and its specific outputs and outcomes. This is where the issue of ethics arises, in the choices we make as to the consequences we desire.

Any archaeological activity can be political in nature and most (if not all) will be. They can operate at a large scale—serving to preserve or destroy a regime or an entire state perhaps, or to provide a reason for a major conflict—or at a small and local scale, encouraging a group of people to work together in a particular way or to understand themselves differently, with consequences for their own self-management and internal relations. It is the contention of the chapter—and indeed the book—that archaeology is inevitably and always a political activity concerned with structures of power and authority, and involving the social, political and economic structures in which we all live. This places heavy burdens upon all archaeologists: we need to be constantly and fully aware of how our work represents and replicates relations of power, the consequences of this for ourselves and others, and the overarching need as a result to adopt an ethical approach to our work. We need always to be alert to questions concerning for whom and in whose interest we conduct our work and to

J. Carman, *Archaeological Practice as Politics and Ethics*,
https://doi.org/10.1007/978-3-031-69828-6_6

recognise the consequences of this for good or ill. We cannot hide behind ideas of professional responsibility as narrowly defined by codes of practice because professionalism goes beyond the mere application of technique (on which more in Chap. 7). As always, however—and as emphasised in Chap. 1—the burden of ethical decision rests upon each of us as an individual. That is why an awareness of the essentially political nature of archaeology is essential.

Archaeology at the Service of Political Entities

As has been noted in other chapters—and is well recognised in the archaeological literature (see e.g. Kohl and Fawcett 1995; Diaz-Andreu and Champion 1996; Diaz-Andreu 2007)—archaeology has a close connection with the development of the nation-state as a political form and nationalism as an ideology. In the nineteenth century, nationalism was a radical rather than a conservative or reactionary doctrine (Kramer 2011), closely associated with liberal, democratic and anti-monarchist ideals. Archaeology—as a study of the distant past through its material remains, which could be linked with distinct ethnicities located in bounded territories—could (and did) provide a scientific basis for claims of national unity grounded in common ethnicity with a deep time dimension. The establishment of new nation states throughout Europe, occasioning also the breaking of multi-ethnic and polyglot empires into separate distinct territories, served—as seen in Chap. 3—to confirm the value of archaeology as a foundation for creating a sense of unity among the newly created national communities. Archaeology was able thereby to gain authority as the science which established the legitimacy but also the inevitable rightness of the nation (defined in terms of ethnic unity) as the proper basis for political organisation and therefore appropriate systems of government (see e.g. arguments in Clark 1939, 1947).

As we have also seen in Chaps. 1, 2, and 5, archaeology was also able to provide the basis for European colonialisation of other territories and the establishment of imperial rule over them by denying any local agency in any monumental pasts and appropriating those pasts to the colonisers as evidence of their natural superiority. McGuire (2008, 14–15) provides some examples of archaeologists directly involved in the imperial adventures of Britain in the Middle East and of the USA in Central America, as well as individuals who rose to high office within nation states. His claims that "in no case has an individual risen to political prominence through or because of their his or her practice of archaeology…. [and] examples warrant mention primarily because they are exceptions in the history of our discipline" (McGuire 2008, 14), although reassuring to those of us opposed to imperialism, are nonetheless somewhat dubious. Although it is true that activities in archaeology can provide good 'cover' for others which are more sinister, a knowledge of local culture and geography—a stock-in-trade of the archaeologist—are vital to the conduct of clandestine political action, especially as a precursor to what may be direct military action. It is no accident that a tool of British archaeology—at home and overseas—has been the maps produced by the Ordnance Survey (originally for military purposes), to the making of which archaeologists have also contributed (Phillips 1959).

A notorious example of the use of archaeology for overtly and evil political purposes has been that of the Nazis in 1930s and 1940s Europe (Arnold 1990, 2006). Arnold (2006) is at pains to show that the ideas dominating Nazi use of archaeology as racist propaganda pre-date the Nazis themselves (and see also Fazioli 2012; Quinn 1994), and Eickhoff (2005) goes further to show a more nuanced perspective both within Germany and outside; in particular Clark—writing for a British audience on the eve of the outbreak of war—is cited as stating that "the main body of German prehistoric research [conducted under the Nazi regime] is not more tainted by bias than it is elsewhere" (Clark 1939, 202; Eickhoff 2005, 74). Clark himself was strongly supportive of archaeology in the nation-building process, citing then recently-established states such as Poland and Ireland and a newly self-governing China (Clark 1939, 190–194). After World War 2 Clark was to modify his position with regard to totalitarian regimes but not significantly so far as the value of archaeology in nation-building was concerned: "coherent nations [will build a] world community.... In a liberal world the operation of national... sentiment can... serve only to enrich society" (Clark 1947, 202).

As noted in previous chapters, archaeologists have become increasingly aware of the essentially political nature and consequences of their work, and as a result Clark's (1939, 1947) happy acceptance of its value in nation-building is increasingly noted and questioned (see e.g. for international examples cited elsewhere in this book Alonso 2013; Dietler 1994; Gänger 2009; Hudson 2021; Mikhailov 2013; Zena 2018). Similarly, and as noted especially in Chap. 5, archaeological work in the heritage sector is essentially nationalist because—despite rhetoric concerning the global nature of heritage—there "is no World Heritage without nation-state sovereignty, and it is nation-states (and experts) that are constituted as the proper actors on the World Heritage stage" (De Cesari 2010, 305). For Gori (2013) this is a good thing since Palestine's inclusion in UNESCO in 2011 allows the construction of a specifically Palestinian archaeological heritage which can be distinguished from Israeli claims, although for Starzmann (2013) the Israeli control of occupied territory and thereby control over archaeological work constitutes a form of colonialism which limits the capacity to do so.

While Israel and Palestine represent two nations disputing their respective pasts, in other territories disputes arise over different visions of the single nation communities occupy: Ratnagar (2004) highlights the case of Ayodhya in India, where claims of destruction of a Hindu temple and its replacement by a mosque concern what it is to be Indian, reflected also in the rise of Hindu nationalist politics; this is archaeology at the service of a politics of separation between Hindu and Muslim communities. A contrasting use of archaeology to unify diverse ethnicities is seen in Korea, where evidence of ethnic diversity is marshalled into a unified scheme so that "all groups and polities in prehistoric and proto-historic Korea must be incorporated into a unilinear evolutionary stage model as a single people" (Park and Wee 2016, 304). Zoh (2020) takes the story of a politicised Korean archaeology further by identifying how the connections of individual government leaders with the pasts of particular sub-regions in the Korean peninsula was mobilised to keep them in power. Rodil (2020) argues for the combination of colonialism and nationalism in the use of archaeology to construct a model of the Philippines as a unified state out of a diverse multicultural background. These diverse examples demonstrate how archaeologists have been implicated in the nationalist endeavour in various ways across the globe.

Others (e.g. Silberman 1982; Meskell 1998; Reid 2002; Bernhardsson 2005; Goode 2007) have noted especially the involvement of archaeology and archaeologists in the initial constructions of nations in troubled areas such as the Middle East and North Africa, especially in the light of more recent developments in those territories. As Pollock (2008, 985) has written in a review of Goode (2007):

> In the run-up to the American-led invasion of Iraq in 2003, archaeologists working in the region played modest but by no means unimportant roles. Some sought out contacts with the [US military] in an effort to ensure that sites and cultural monuments be spared damage. Others contributed to narratives that linked the ancient civilizations of Mesopotamia to the modern-day West. Still others spoke out against the invasion.... For many archaeologists who worked the Middle East, the invasion of Iraq confirmed—although not for the time—the extent to which contemporary politics could affect their work.

Whether archaeologists should or should not work with the military in Iraq and Afghanistan—both subject to military action by Western countries and their allies in an effort to remove the regimes in power there—became an issue of debate within the field. Several publications (see e.g. Starzmann et al. 2008; Stone 2011; Stone and Bajjaly 2008; Rush 2010) specifically addressed the issue and it was also raised at a debate at the 2008 World Archaeological Congress in Dublin and further at an Intercongress held in Vienna in 2010, although never quite resolved. Arguments turn on issues of where archaeological responsibility lies: if to the archaeological record, then the role is primarily to ensure the protection of sites and material from damage by military action, deliberate or accidental; but if more widely to the communities whose heritage is threatened by military action, then it may be to avoid giving any assistance to military forces of any kind. Reports of the involvement of anthropologists in the Human Terrain 'hearts and minds' operations—designed to encourage support for and assistance to invading and occupying forces in Iraq and Afghanistan—were worrying to many as a rejection of scholarly detachment from particular ideologies (Perugini 2008).

Perring and van der Linde (2009, 201) point out that "Archaeologists [involved] cannot place themselves above and beyond the process of conflict. We are not neutral observers but protagonists: complicit and partisan". This is because any involvement—as advisors in preliminary planning stages, during conflict as advisors to participants, or afterwards in post-conflict reconstruction projects—is a matter of political choice which affects both the conduct and outcome of the conflict; and to choose no involvement is also to make a political decision with consequences for participants and the archaeological heritage. They make a particular point about reconstruction efforts—most commonly approached as apolitical and neutral exercises in heritage conservation and repair—as "a continuation of conflict by other means, and show how archaeologists are inevitably drawn into opposing camps when the archaeological past is itself the coin of dispute" (Perring and van der Linde 2009, 204). They argue ultimately for "sound and scientific method, the practice and teaching of which is usually the main justification for a foreign involvement, and a self-critical approach to the uses of archaeological science that is informed by the post-processual debate.... [so that] debate over the contemporary roles that we can find for the tangible and intangible heritage can also be used to redirect hostilities into areas where negotiated outcomes can be achieved" and peace built (Perring and van der Linde 2009, 204–205). In thus emphasising the ethical responsibilities of archaeologists in these contexts, the link is made to an underlying and continuous theme of this book. What is nonetheless interesting is that they offer only the option of working within existing structures: it seems we are always forced to support the institutions of the political *status quo* even as we work to overcome them or avoid contamination by them.

Luke and Kersel (2013) demonstrate in specific detail the various ways in which archaeology represents or is used by US governments as a tool for international diplomacy. The various American Schools located in other countries as centres for research and education, liaison with governments over permissions to carry out research, the provision of finance to support cultural activity and development, and the legal and other enforcement of preservation measures, are all designed to promote the US as a friendly nation and encourage others to accept US policy as legitimate. As they explain, as "archaeologists move between the upper echelons of [foreign] governments and the local communities [in undertaking their work, and].... whether they recognise it or not, they are very much embedded agents of U.S. soft power and unofficial, long-term cultural diplomacy" (Luke and Kersel 2013, 13). As they emphasise, US government "programs [sic] and policies... are most often highlighted as apolitical" (Luke and Kersel 2013, 16) but in practice are "reactive, focused on contemporary U.S. interests and security" (Luke and Kersel 2013, 18). It is of course not only the USA that indulges in archaeology as diplomacy: Australia (Clarke 2017) and China (Storozum and Li 2020) are also especially noted as being engaged in such activities, and heritage more generally has been the subject of quite widespread discussion as diplomacy (see e.g. Winter 2015; Todorović 2022; Lähdesmäki and Čeginskas 2022). In all cases, archaeology as a branch of foreign relations acts as a servant of established political structures for the discipline's own advantage: it provides opportunities for and access to research, the chance of forming international collaborations, and provides states with limited resources an extra capacity to protect its archaeological resource. However, it also means that archaeologists who—wittingly or otherwise—are engaged in the exercise of such soft power relationships confirm the authority of the powerful over the weaker, and any claims to act in the name of ameliorating the North-South global divide are more likely to result in exacerbating it by confirming existing power disparities.

A similar exercise of authority can be seen within nation states as well as between them, as argued by Karl (2020) for German-speaking countries in Europe. He quotes Pollak (2011, 227) on the "lack of a debate of the foundations of archaeological heritage management… due to a general consensus of all participating experts" who—as Karl (2020, 151) argues—are held to be those who deliver 'truths' that cannot be challenged and whose pronouncements about the archaeological heritage are automatically considered to represent the public interest. The latter is of course the kind of public service archaeology discussed in the last chapter, which operates at a general level of 'public welfare' detached from specific benefits accruing to actual human communities (see also Merriman 2004, 1–2; Carman 2002, 97–112). Karl's (2020) argument is of course similar to—and indeed in part derived from—Smith's (2006) notion of the 'authorised heritage discourse' and the politics of archaeology as set out in her account of the relationship between archaeological theory (especially in the guise of American processualism) and the application of expert knowledge as a form of power (Smith 2004). As Karl (2020, 152) sums up his argument, "archaeologists reign absolute [in Germany and Austria] because we have managed to preserve, not so much the archaeology, but a model of public administration that does not serve the actual public interest in archaeology, but primarily our self-interest". As others have also pointed out (e.g. Smith and Waterton 2009), archaeologists themselves constitute an interest group in their own right (and see Chap. 7), so their statements and actions in respect of the archaeological heritage cannot be considered entirely disinterested. This is of course not entirely the use of archaeology to serve others' interests but it relocates archaeology in the realm of a political force in its own right—the topic of the next section.

Archaeology as an Agent of Governance

Karl (2020, 145) demonstrates the "hierarchical order of authority in the Germanophone archaeological heritage discourse" whereby those associated with the government are most trusted, closely followed by all those expressly professionally qualified. The least respected and trusted are those who are not officially qualified—e.g. amateurs and metal detectorists—or behave towards archaeology in a manner not acceptable to the professional archaeological community, such as illegal excavators and developers. While Karl is limiting his discussion to the context of Germany and Austria, it is not so different a position from that of Smith (2004) in her argument about archaeology as a mechanism of governance in Australia and the USA. Smith's (2004) critique of archaeology as governance relates especially to relations with Indigenous peoples who, she argues, are classified by reference to their material culture as interpreted by archaeologists. She points out how it is the discourse of archaeology as a scholarly activity which is translated into managerial action as cultural resource management (the topic of Chap. 5) and which in turn "helps to maintain the intellectual legitimacy of the discipline in the eyes of bureaucrats and governments" (Smith 2004, 195) and "is encapsulated in the position the discipline occupies as a technology of government" (Smith 2004, 196).

The close relationship between archaeology and the nation state is a theme of this book as well as the wider literature of the field but the closer relationship between archaeologists and specific acts of government has a longer history. Elsewhere (see e.g. Carman 1996, 67–96; 1997, 2005, 2010, 2015, 15–23), I have discussed the role of an emerging archaeology in policy-making in Britain that goes beyond a concern for the past. Those involved in the process—especially John Lubbock, later Lord Avebury—were not exclusively concerned with the past, but to put their understanding of the past at the service of their contemporary politics. As a Liberal politician, Lubbock was as concerned with contemporary social change as that in the past, and was also concerned with the position of Ireland within the United Kingdom. Biographers such as Patton (2007) make it clear how Lubbock's separate interests in science, business and its conduct, public welfare, education, personal behaviours and politics were all related to and driven by ideas about social evolution and fears that civilisation would be overborne by the savagery (as he saw it) of those unable to govern their passions, leading to the horrors of violent revolution. In relation to Ireland, Lubbock's understanding of the past comes particularly to the fore, especially in debates with his party leader, William Gladstone, about the distribution of ethnic groups through the British Isles. For Gladstone, the British Isles was inhabited by four separate nations: the English, the Scottish, the Welsh and the Irish, justifying a drive to grant Ireland a measure of devolved government; for Lubbock, a number of different ethnicities—the Celtic, the Anglo-Saxon, the Norse—were distributed identically in both main islands, with the Celts to the west, the Norse in north and east, and the Anglo-Saxon to the south and east, justifying a full political union of Ireland with mainland Britain. A concern to protect archaeological monuments ultimately—after a ten-year Parliamentary campaign—resulted in legislation to achieve this, but the real purpose was to establish the ultimate truth of gradual social change by evolutionary process as evidenced (as Lubbock and his allies believed) by remains from the prehistoric past, all as the foundation for a public education programme based around science. This was British Liberal party politics redefined as science and given a deep-time dimension.

McGuire's (2008, 14) claim cited earlier of the general innocence of archaeology in relation to political office does not include Lubbock as an example, but Lubbock—although prominent as a Unionist on Irish issues, in regulating new and emerging professions and other commercial activity, in public welfare and in educational matters—never rose to office. However others prominent in developing the study of the past through its material remains were also government officials. In particular, following the collapse of the Ottoman Empire as a result of defeat in World War I, Gertrude Bell was appointed Oriental Secretary to the British High Commission in Iraq where she was responsible for establishing the boundaries of the new state and appointing its new monarch (Courtney 1926, 660). Bell's close ally T. E. Lawrence (of Arabia) was one of those providing survey data of the Sanai Peninsular in advance of British military action in the area in World War I and was also active in promoting Arab interests at the Paris peace conference. In case it should be thought that only the British carried out such nefarious activities, McGuire (2008, 14) cites the case of Sylvanus Morley whose investigations of Mayan sites in the Yucatán were cover to negotiate with rebel Mayan leaders for their support of US interests and whose survey work on the coast of Central America during World War I was also to identify German submarine bases (Price 2006).

Arguably more legitimately, engagement in the archaeology of others' lands as a sideline to government work is notable throughout the nineteenth and into the twentieth century. Lord Elgin's acquisition of the Parthenon (or if you prefer Elgin) Marbles between 1801 and 1812 from the Turkish authorities in Athens was secondary to his position as Ambassador to the Sublime Porte (the Ottoman Empire) (Greenfield 1989, 61). In Mesopotamia, Paul Emile Botta, French consul at Mosul, was active in excavating sites from 1842 (Greenfield 1989, 108) and Sir Austen Henry Layard, attached to the British ambassador's staff, other sites from 1847, but Layard was also engaged in border disputes; Layard went on to a political career founded upon his fame as the excavator of Nineveh and rose to under-Secretary of State for Foreign Affairs (Goldsmid 1894, 372–373). In India, Lord Curzon, appointed as Imperial Viceroy, was active in studying and preserving Hindu architecture (Thapar 1984, 65) and later rose to Foreign Secretary. Not to be outdone by the imperial nations, Papal expeditions for missionary purposes in 1925 acquired significant numbers of objects from across the globe (Greenfield 1989, 109). In all these cases, archaeology (not always ethically conducted) was secondary to the purpose of the appointment; but it was the status of the position in government that also provided the individuals involved with the requisite authority to investigate, excavate and remove material. In the case of Layard—whose appointment to the ambassadorial staff was always unofficial—it was his notoriety as an excavator that fuelled his later career—a challenge (albeit only one) to McGuire's claim that none achieved office through archaeology.

The concerns of those such as Lubbock was always with their own people: although those of Celtic ancestry were recognised as distinguishable from others in his interpretation of the British past, they were not to be treated separately as an Indigenous population. It is however in relation to Indigenous communities that issues of archaeology as a form of governance particularly arise. Citing the collection by archaeologists and anthropologists of Canadian Indigenous peoples' cultural objects and human remains, Steeves (2015, 123) emphasises how such "items have been used to interpret and define Indigenous histories and peoples… without any input from Indigenous peoples or communities". She goes on to cite cases such as that of Kennewick Man (also cited by Smith 2004, 161–172) as a trend of "exclusion and control" of the Indigenous over their own pasts and indeed identities (Steeves 2015, 135). Also writing on Canada, Dent (2017, 148) demonstrates how collaborative arrangements between state agencies and officials and Indigenous groups nonetheless do not "fundamentally change the underlying state-centric structures and processes within heritage management" but only create the—often unfulfilled—possibility of such change to the advantage of Indigenous communities. As it is explained, these collaborative arrangements "are functionally about providing direction and clarity with respect to disposition of material heritage as opposed to any changes in authority over how archaeological and other tangible and intangible heritage resources are managed" (Dent 2017, 142), thereby maintaining the state authority over archaeology and thereby also over the Indigenous communities whose heritage it is. Such control inevitably benefits professional archaeologists, who—as Steeves (2015) points out for North America—may have little or no training in Indigenous cultures and ways of life but who are the ones who can nonetheless decide the fate of Indigenous culture.

Huchings and Dent (2017) list the factors that closely relate archaeology to government, a relationship they characterise as the "Archaeology-State Nexus" (Hutchings and Dent 2017, 3). Their focus is upon North America, a territory generally regarded as one resistant to State control and yet even here:

- virtually all archaeology performed is state-sanctioned, as cultural resource management (CRM) now constitutes upwards of 90 per cent of the practice....
- virtually all academic archaeology is state-sanctioned, as most work is done in state institutions with state funding;
- virtually all academic archaeologists are state agents, as most are employed by and thus responsible to the state;
- virtually all professional archaeologists are state agents insofar as most are paid to enforce state heritage regimes (i.e. compliance archaeology, or CRM); and
- the primary focus of archaeology is Indigenous heritage—thus living Indigenous people (Hutchings and Dent 2017, 3).

Hutchings and La Salle (2017) in the same collection go on to accuse archaeology of "State Heritage Crime"—that is crime committed by (rather than against) the state—because archaeologists' failures to redress power imbalances against subaltern communities are the product of archaeology "working as designed" (Hutchings and La Salle 2017, 68). As they point out, archaeologists themselves choose to disavow any engagement with peoples or places occupied by force (WAC 2013). Since they define archaeology specifically in terms of governance—as "a technology of government designed to control Indigenous heritage and thus living Indigenous people and their resources" (Hutchings and La Salle 2017, 68)—this failure, despite archaeologists' own rhetoric as exemplified by WAC, "is criminal" because "it contravenes international treaties, violates Aboriginal law and custom, and causes significant social harm to the communities connected to the heritage at stake" (Hutchings and La Salle 2017, 78).

Taking a more international approach to understanding how archaeology governs communities, Lafrenz Samuels (2009) discusses the intersections of archaeology and economic development. She accuses archaeologists of limiting the discourse of the economic value of remains to issues of "cultural property, the antiquities market, and the costs of running cultural resource management operations" and thereby ignoring "the profits of 'enclave' heritage tourism being funnelled out of communities to international corporations..., the emphasis on nonmaterial 'universal' aspects of archaeology such as identity and rights at World Heritage sites so as to mystify the appropriation of property rights... and heritage policy interventions of international organizations that reproduce the status quo of global structural inequalities through appeals to expertise" (Lafrenz Samuels 2009, 70). She argues that efforts to reduce poverty allow "transnational control over resources, finances, and social policy, via the production of 'expert' knowledge, particularly in the field of environmental management, and its offshoot heritage management" (Lafrenz Samuels 2009, 71). There is a connection here with the mobilisation of archaeology as a diplomatic tool cited earlier (Luke and Kersel 2013) since poverty reduction programmes are also often tied to efforts to introduce political reforms, especially in areas where representative democracy is rare, such as the Middle East and North Africa (Lafrenz Samuels 2009, 73).

The justifications for archaeology as a provider of benefits are rarely couched in terms of archaeology as an arm of government (see e.g. British Academy 2016), but it is notable in the rise in interest over past decades in proving the value of our work in support of specific areas of policy. Rypkema et al. (2011, 3–4) specifically list those areas where the benefits of historic preservation (one name for the topic of Chap. 5) can be identified in economic terms:

- jobs and household income—by the provision of paid employment
- rises in property values—as a direct result of identification as historic in themselves or located within historic districts
- heritage tourism—as separate from other forms of tourism locally
- environmental benefits—the contribution of historic preservation to broader 'sustainable development,' 'Smart Growth,' 'energy conservation,' and environmentally-sensitive or 'green' community planning
- 'downtown revitalisation'—the role of historic preservation and downtown, commercial district revitalization.

Similarly in the UK, APPG (2011, 5–6) identified the economic values of heritage as the commercial value of heritage properties, growth in local business activity, tourism, revitalisation of town centres and rural settlements, and the availability of construction and related employment. Although categorised as economic measures, these of course relate directly to areas of direct concern to governments in the developed global North: overall economic strength and growth, business viability, foreign income provision, public welfare, and of course widespread employment opportunities. Compared to other economic and commercial sectors archaeology may be relatively small, but the provision of heritage resources generally contributes to taxation receipts and population security and general acceptance of political authority. In respect of the latter, the APPG report also notes additional values accruing from heritage: regenerational value—a key element in establishing a greater degree of equivalence between regions; environmental value—especially with regard to reducing greenhouse emissions; and social value—including creation and maintaining firm senses of identity and community cohesion, together with benefitting mental health (APPG 2011). All of these work to the advantage of government in its widest sense by ensuring allegiance—or at least limited public challenge—to the established order.

Archaeology as a Response to Power Relations

There are frequently one of two modes of archaeological response to their involvement in the realm of politics. One is passive and is that of 'the archaeologist as victim'; the other is active and is that of the protester against misuse of the past.

While the tone of voice differentiates the two positions, the basis of each is the same: that others misuse or misapply the products of archaeology for their own purposes. Taking such a position serves a useful purpose—it relieves the archaeologist of responsibility for the manner in which material and sites and their interpretations are presented to others and for the consequences—intended or otherwise—of the work of archaeologists. As Meskell (2002, 293) notes "Part of our problem rests with the illusion that the subjects of our research are dead and buried, literally, and that our 'scientific' research goals are paramount" which supposedly detaches us from the arena of politics and puts the blame for any use of our interpretations elsewhere. As Arnold, in arguing for a sensitivity to the forces that affect the discipline has recently put it, "archaeologists and the cultural heritage they are professionally trained to recover, interpret and protect are still pawns in power struggles between entities that frequently have no scholarly interest in the past apart from its utility in underwriting their political or economic agendas" (Arnold 2023, 17). In adopting such a position no mention is made of the effects of archaeological activity: it assumes the automatic legitimacy of what we do and therefore its outcomes. The issue perhaps is not, however, what others do to us and our work: but how we should respond.

This is not to deny the effect of externally-applied political forces upon archaeology but the archaeological response to them emphasises the essentially interested nature of archaeology as a discipline. Aitchison (2009) commented upon the consequences for archaeology of the 2008 financial crisis then affecting the world, noting its differing effects in the UK, Ireland and Japan: his call was for archaeology as a sector "to demonstrate flexibility and adaptability to cope with and ultimately to profit from the new circumstances that we will find ourselves in" (Aitchison 2009). A few years later, the effect of so-called 'Brexit' (the departure of the United Kingdom from the European Union) caused widespread dismay in European countries, neatly expressed by Schlanger before any agreement about actual separation had been reached:

> not only is Brexit set to have detrimental material impacts on archaeological research and heritage management for years to come, affecting personnel, funding and legislation, it has already begun… to blow its dark clouds of suspicion and despondency (Schlanger 2017, 9).

His answer is to ask: "'what can we do, together?' How to stand shoulder to shoulder—academic and commercial archaeologists alike, not forgetting the museum sector, culture and the arts—to ensure that statuary obligations are upheld, that heritage management remains viable, that scientific expertise, epistemic authority and moral integrity are maintained; and indeed that the past we discover, produce and disseminate does not" become a tool "for 'leveraging' the image of post-Brexit Britain"? (Schlanger 2018, 1668). The concern here was especially the risk of the development of a narrowly nationalist and chauvinist archaeology in Britain once politically and economically separated from the rest of Europe: one that would damage the opportunities of individual archaeologists to work in countries other than their own, limit the funding for archaeological projects, especially those investigating cross-border phenomena, and prevent cooperation between international colleagues. These are concrete concerns relating directly to archaeological practice but simultaneously reflecting theoretical issues around identity construction, providing a neat example of the lack of any real theory/practice divide.

An issue deriving from the previous section on governance is the diverse nature of branches of government involved in archaeology and how they act upon the discipline, as nicely demonstrated by Bigambo (2023) for Tanzania and Ndlovu (2016) for South Africa. For Tanzania, responsibility for archaeology is divided between the "Antiquities Division from the Ministry of Natural Resources and Tourism [and] the National Museum of Tanzania" each of which has control over different aspects of work and different classes of material, none of which prevents duplication of and conflicts over jurisdiction, while some classes of heritage (such as intangible heritage) are excluded from any oversight because excluded from the concerns of either (Bigambo 2023). In South Africa, the government agencies cited as granting the necessary licences for mining in a World Heritage Site included "the Department of Minerals and Energy (DME, approving department); Department of Environmental Affairs (DEA, approval of environmental authorisations), Department of Arts and Culture (DAC, commenting department—through its agency—the South African Heritage Resources Agency) and the Department of Water Affairs (DWA, approval of the water licence)" (Ndlovu 2016, 287). Here, governance acted to the detriment of the historic environment, as it has elsewhere in the world. As noted in Chap. 3, Ronayne (2007) reports on the extremely detrimental effects on cultural heritage but more importantly on communities of the construction of a dam in Turkey: the point here is that it was express policy decisions and government action that directly worked against the interests of both archaeological and living culture in the name of providing other kinds of benefits, triggering a more active response. In this case, the benefits accrued to the population of one country but were also at the expense of the populations of others downstream from the dam construction. In cases such as these, government organisation and policy are clearly antithetical to the interests of archaeology, despite the general agreement across the world that governments are responsible for the preservation of their national patrimony (as required by e.g. UNESCO 1972).

A more seriously egregious impact of government on archaeology in the recent past has been reported upon by Papoli-Yazdi (2023) who—drawing upon a discarded notebook found in a rubbish bin—documents the oppression of archaeologists and other scholars during the Islamic so-called 'Cultural Revolution' in Iran in which "thousands of students and professors were framed, tried, fired, and arrested" (Papoli-Yazdi 2023) as a result of being reported to the authorities by the Islamic Association of Students. The document reveals how "the field of archaeology was reduced to a nationalistic governmental tool in the hands of the powerful people" and leads to the comment that:

> It is important to keep in mind that many nationalist archaeologists fulfill their careers by publishing and studying abroad.... These archaeologists use the credit taken from their international studies and publication to prove that their interpretation of the archaeological evidence [in support of government-sponsored myths their home country] is the only 'correct' understanding of the material and the 'pure truth' of the past. Documents such as [the

> one discussed] remind us of the importance of revision of professional ethics in terms of being careful about the projects we support and the articles we help to be published as well as the ideas we give a chance to be circulated…. We, the archaeologists, need to raise our awareness about the dark side of the field we are experts in. Otherwise, archaeology may give a hand to the dictators who intend to manipulate the history as well as the destiny of dissident archaeologists and many other people (Papoli-Yazdi 2023).

This is at once a good example of the 'archaeologist as victim' discourse but also serves as a warning to us all that because of our specific area of interest we are subject to targeting by those who wish to control understandings of pasts. It is a recognition of the significance of archaeology as a field of activity: archaeologists are not targeted by malignant regimes for no reason; it is because what we do matters.

In contrast with the idea that political forces act upon archaeology to its detriment, and distinct from the involvement of archaeology as an agent of governance, is the recognition that archaeology has an active role to play in constructing the contemporary world. Smith (2017) has argued that the rise of archaeological studies of the contemporary past (as discussed in Chap. 3) "can help us attain a deeper understanding of the critical social and political issues of the day, such as racism, discrimination and mass emigration, and to identify material ways to reinforce social cohesion" (Smith 2017, 26). She emphasises a number of developments in archaeology over recent decades—some referenced already elsewhere in this book—including decolonising academic study, which is "characterised by a concern with rethinking colonial histories…, deconstructing power relations between Indigenous peoples and archaeologists…, shared histories… and shaping archaeological practice so that it benefits Indigenous communities" (Smith 2017, 9–10); community archaeologies (discussed in Chaps. 3, 4, and 7) (Smith 2017, 13–18); and archaeologies for social justice, concerning "issues of equity and human rights… and wider concerns about structural violence" (Smith 2017, 19; and see Little (2007) on 'restorative justice' mentioned in Chap. 1). Smith's is a call to action rather than action in itself—and one particularly directed towards archaeologists, especially those outside the Anglophone world: as she puts it, there is too much of "an equation of 'us' [as archaeologists] with the 'West' and an over-representation of Anglophone scholars…. If archaeology… is to truly flourish…, it will need to be embraced by the perspectives of European, Asian, South American and African scholars" (Smith 2017, 26). She advocates for a definition of archaeology that goes beyond a concern for the study of the past to have "a clear social and political purpose [and] its own role to play in sculpting the world around us" (Smith 2017, 35).

It is no accident that calls such as Smith's (2017) were made primarily to the discipline because in order to exert any influence on the contemporary world, archaeology was deemed to need first to reform itself to do so. As Shepherd (2002, 190) explained for the lack of discussion of the relations between archaeology and politics in an African context, "the reasons have to do with the new archaeology [of the 1960s and 1970s which in] its concern to remake archaeology as a positivist science, had specifically rejected the notion that knowledge is constructed within cultural, political, and economic contexts". As European investigators of African pasts followed on the heels of conquerors and colonisers, the discipline was engaged in "actively constructing archaeological pasts over and against local and indigenous conceptions" (Shepherd 2002, 192), a process that has since been exacerbated by post-independence debates on the interpretation of archaeological findings and the implementation of heritage management processes imported from the global North (as also noted by e.g. Bigambo 2023 for Tanzania). Shepherd concluded that the "key issue for practitioners in Africa remains the manner in which they negotiate a place in a globalized field of archaeological pursuit.... The political challenge for archaeologists in Africa lies in posing the kinds of questions, and developing the forms of practice and the kinds of understandings of our own place and purpose, which begin to tilt this relation in the other direction" (Shepherd 2002, 205). It is clear that Shepherd's (2002) challenge to African archaeologists required renewal a decade and a half later, and to a wider global audience (Smith 2017).

Much of the discussion such as Shepherd's (2002) and Smith's (2017) is specifically focused upon issues of race and ethnicity, issues that arise particularly in the context of territories where a subaltern ethnic group—Indigenous or otherwise—suffers at the hands of a dominant population. In South Africa the legacy of apartheid (specifically referenced by Shepherd 2002, 197–199) was of a hierarchy of ethnicities with the Indigenous Black African population as the least advantaged; in the countries of Australasia, North and South America, Indigenous populations were subordinated by incoming Europeans; also in the Americas the legacy of enslavement created an unwillingly diasporan African community similarly disadvantaged; in northern Europe the Sámi suffered dominance by the peoples of Scandinavia; and the Roma of Central Europe continue to be ignored as a distinct community in the countries of that region (MacDonald 2013). Lippert (2006) makes a case for the role of archaeological organisations—specifically the World Archaeological Congress—in supporting Indigenous communities, while Cuddy and Leone (2008) emphasise the role of archaeologists in engaging with the material culture of African Americans. Funari and de Carvalho (2011) take a wider perspective on how archaeologists engage with community identities in Brazil—making no distinction between those of Indigenous, African or European descent—although noting it is a factor together with others such as class and gender. The distinction between areas of the globe relates to differing ideologies relating to identity construction: differences of ethnicity are major factors affecting all aspects of life in the USA especially, but elsewhere are sought to be overcome rather than emphasised. In large parts of Asia and Africa, differences of ethnicity are subsumed within an overarching nationalist ideology which is designed to avoid the recognition of difference (e.g. Carman 2023). The archaeological responses to inequalities based upon issues of race therefore differ from territory to territory, which is one reason for the continuing lack of fully global response to Shepherd's (2002) and Smith's (2017) calls.

There is of course a clear distinction between a concern for issues of inequality based upon ethnicity and race in the present, and research to understand how such processes operated in the past. Much of the archaeological response to such issues is inevitably in terms of research into the past to inform the present, as exemplified in North America by a number of scholars (see e.g. Cuddy and Leone 2008; McDonald et al. 2006; Ferguson 1991; Orser 1991; Bower 1991), rather than direct action in the present (the topic of the next section). Distinctions between classes of people based upon ethnicity is especially a topic for the archaeology of more recent times, following the European colonisation of others' territories (e.g. for the sixteenth and subsequent centuries in the Americas, Gasco 2005; Lightfoot 2005). As discussed in Chap. 3, such work also addresses the lives of others affected by the spread of European colonial and imperial activity, such as Karsken's (2001) study of Chinese immigrant communities in nineteenth century Sydney, Australia. As indicated above and in Chap. 3, there is a particular focus on Indigenous peoples, especially in the Americas (see e.g. Rubertone 2008; Gnecco and Ayala 2011) and Australasia (Smith and Wobst 2005), but increasingly also in far North Europe (Harlin 2019) and elsewhere (e.g. Pratap 2009; Eze-Uzomaka 2000; Varouchakis 2017). As also mentioned in Chap. 3, others (see e.g. McGuire and Paynter 1991; Mayne and Murray 2001; Mrozowski 2006) focus on class relations, especially in urban settings. This is part of the widening concern of archaeologists for understanding the more recent past but also a reflection of the rise in social awareness of the Anglophone discipline, as recognised by those such as Smith (2017) cited earlier.

While studies of ethnicity and race relations have been especially prominent among archaeologists, equally significant have been developments in the appreciation of gender as a factor in constructing the social order. Hays-Gilpin (2000, 89) claimed that "Archaeologists have been slow to embrace feminist scholarship" although she demonstrates that the first publications relating to feminism to archaeology emerged in the early 1980s (predating by a few years emergent concerns for relations with Indigenous and other subaltern groups) and of course women had played significant roles in the field since its emergence in the nineteenth century. The review goes on to highlight the many and diverse ways that research into sex and gender relationships has taken: a search for the origins of the oppression of women by male dominated societies; studies of the sources of different kinds of power wielded by different sections of society; a search for examples of gender-egalitarian societies; a rejection of the simplistic sex-gender dichotomy; and the 'queering' of categories to include studies of the range of possible sexualities together with the constructions of other genders—including the masculine and juvenile (Hays-Gilpin 2000, 98–101). The more recent emergence of intersectional studies (referenced in Chap. 3) demonstrate how categories such as ethnicity, gender, class and sexuality, together with such factors as "religious affiliation, profession…, citizenship…, health and ability,… educational levels, and age" work together to construct social relations (Spencer-Wood and Trunzo 2022, 23). The crucial factor in encouraging all these developments has been the concern of feminist theory with disparities of power and authority, in large part derived from the experience of women in archaeology, as discussed especially in Chap. 2.

This focus on identity construction is primarily an effort to understand how our own world has come to be the way it is and in part also to find models for alternatives in the past. They are—as mentioned in Chap. 4 and indeed frequently acknowledged by their practitioners (e.g. Potter 1994)—a critical approach to the past rather than a direct intervention in the present. They also require the exercise of imagination, which has been offered as a vital skill in archaeological interpretation (van Helden and Witcher 2019). The topics addressed in the study of gender, ethnicity, class etc. makes archaeology more than the study of the past but also a contribution to the understanding of ourselves, and providing ways towards futures, a topic addressed in Chap. 3.

Activist Archaeologies

There is a narrow distinction between attempts to define alternatives to the present and activism designed to bring them about, inevitably arbitrarily drawn. I have chosen to make the distinction on the basis of what is written to describe the nature of power structures, and direct action in various forms (the former then becomes a reaction; and the latter a direct challenge). It is of course the case that many cited as challengers to power are also activists, further emphasising the arbitrariness of the division. Nonetheless, it is a fair one: my own work can count as a challenge since I am not noted for taking my opinions to the street nor working with others to threaten authority and accepted ideology. Others are bolder and need to be recognised as such. As mentioned in Chaps. 1 and 4, Sabloff (2008) specifically advocates 'action archaeology' in a number of key areas where archaeological understanding of pasts can make a significant contribution: creating a sustainable environment; ending war; making cities more liveable; and preserving cultural heritage. This puts Sabloff's book somewhere between my two poles of response versus action: it provides not only commentary upon which others can draw, but also maps out specific practice that can be adopted. Similarly, Stottman (2011) asks "Can archaeologists change the world?" and clearly answers in the affirmative: in this case it is through mechanisms and examples of community engagement. The various forms of community engagement have been specifically discussed in Chap. 4 and will be further addressed in Chap. 7, but here the focus shifts to how those engagements represent involvement in political action and the kinds of politics engaged in. The latter depend on where the activism is taking place and the issues addressed: as mentioned above, in territories colonised by Europeans the most prominent issues are very often those of race relations; in other areas they concern inequalities of wealth and access to resources; elsewhere again they concern overcoming oppressive rule of some kind. The choices are also made by individual archaeologists and relate to issues that affect them in particular ways.

One way in which archaeologists engage directly with contemporary issues draws specifically upon their skills and practices to investigate the relationship of material culture with lived experience. An example is that of Smith (2012), who drew upon ideas about the spatiality of material culture to explore how asylum seekers in Ireland experienced the 'accommodation centres' in which they were placed and the methods of control under which they lived—including separation from local communities, tightly regulated limitations on activities they were able to perform, close surveillance in communal areas, and shared and crowded 'private' spaces such as bedrooms (Smith 2012, 359). The specific social engagement of the work—which qualifies it here as an example of activism—involved dissemination of the findings to relevant organisations engaged with immigrants, to develop "both government sponsored and community-based integration policies and programs in Ireland to work towards better inclusion of this marginalized group in Irish society" (Smith 2012, 360–361). In similar vein, Zimmerman and Welch (2011) have addressed issues among the homeless in American cities to provide specific insights into life as a homeless person, pointing out that an archaeology 'of the last ten minutes' "works at a boundary between past and present with a fundamental understanding that humans constantly create archaeological sites" (Zimmerman and Welch 2011, 82) and thereby contributing to the understanding of those who seek to help who are 'invisible' on the streets. Quite independently, Kiddey (2017), also cited in Chap. 3, conducted similar but more extensive work among the homeless in Britain, advocating an 'ethic of care' for such work, suggesting that "participatory cultural heritage work might explicitly set out to function as a therapeutic social intervention with marginalized communities" (Kiddey 2017, 160).

Others engage directly with disadvantaged communities themselves to bring about changes both within the discipline and the rest of the world. Schaepe et al. (2017), cited in Chap. 3, emphasise the value of working with Indigenous communities to reconnect their presents and pasts, and the people to their environment. Similarly, Laluk et al. (2022, 678) promote a form of "sovereignty-based social justice framework" whereby authority is given to Native American communities and Indigenous archaeologists to address persistent injustice. As mentioned above and in Chap. 3, Ronayne (2007) built an alliance of archaeologists with a local community to resist the building of a dam which would have destroyed archaeology and the communities whose heritage it was. Green et al. (2021a, b) appropriated the idea of 'archaeology as a public good' (discussed in Chap. 5) to redefine it in terms of "how archaeology contributes profoundly to a contemporary community's coherence, vitality, and its sustainability" (Green et al. 2021a, 13) and introduce a series of articles each of which outlines a project designed to achieve this aim. Green et al. (2021b) describe how work in southeastern Ireland will create a comprehensive preservation plan, leading to the construction of a historic landmarks trail which in turn will encourage ecotourism and action to mitigate the effects of climate change on a fragile coastline. In the same collection, Ralph et al. (2021) describe a project to provide employment and therefore reliable income to an Aboriginal community in the Northern Territory of Australia based around the recording of the unmarked graves of the Indigenous inhabitants. Rankin and Gaulton (2021) discuss work in eastern Canada which aimed to develop tourist resources and rebuild disadvantaged fishing communities in a remote region.

These examples—and others of community engagement cited in Chap. 3 and elsewhere in this book—mostly work within the accepted structures of conventional political discourse, helping the disadvantaged by allowing them to benefit from established methods of living in the world. Taking an explicitly Marxist perspective,

McGuire (2008, 16) argues that "archaeology is politics by other means" because "archaeological practice produces ideology"—mostly and most usually by supporting the current *status quo,* as argued above specifically in relation to archaeologists' engagements with conflicts (and see Chap. 5 on archaeology as public service, most commonly in the service of the state). McGuire argues that archaeology "is a weak weapon for political action, because it cannot be wielded directly in the struggles over land, life, liberty, and wealth that drive the political process", but his assertion is challenged by the work of others who engage very directly in precisely those things through various kinds of emancipatory archaeology (e.g. as cited in Chap. 3 Ronayne 2007; McDavid 2004; Saunders 2002; Crossland 2002; Ballbè et al. 2007; Meskell and van Damme 2008; Kiddey and Schofield 2011). The point of difference from McGuire—if indeed it is a serious one since he also advocates an emancipatory archaeology—is that these examples all work with particular communities in specific contexts. They challenge the way things are by acting in concert with others at the local level addressing very specific situations.

Challenges specifically to the role of archaeology in providing ongoing support to capitalism are not exclusive to McGuire (2008)—see also Tilley (1989) and Hamilakis and Duke (2007)—and an underlying point of the public archaeology programme at Annapolis was specifically to demonstrate to the non-specialist audience how modern capitalism emerged and that the world need not have developed the way that it did (Potter 1994). Others have also emphasised capitalism as a theme of investigation (e.g. Johnson 1996), especially to provide a basis for an academic critique. But anti-capitalist radicalism is not restricted to those who subscribe to Marxism. As Flexner and Gonzalez-Tennant (2018, 213) outline, "many archaeologists have come to embrace an interest in anarchism in some form". Much engagement with anarchist ideas relate to identifying and understanding non-hierarchical and egalitarian social forms, ethnographic, ancient and modern, and how resistance is mounted against authority. However, "archaeologists also engage with anarchist praxis on some level [which] means working to eschew or undermine certain forms of authority, including those we ourselves might be given" (Flexner and Gonzalez-Tennant 2018, 215). As Eddisford and Morgan (2018) demonstrate, as 'neutral' a fieldwork technique as single context recording can be mobilised to represent anarchist practice because it represents the distribution of knowledge production on archaeological sites, relying upon the 'natural' authority of expertise and experience rather than imposed hierarchical structures. a distinction especially drawn from the writings of Bakunin (Eddisford and Morgan 2018, 150). They emphasise also the inevitability of the "trench-side chat" in the production of interpretation and the inevitability of co-working (a point made elsewhere in this book and revisited in Chap. 7) as an example of the kind of voluntary co-operation that lies at the heart of anarchist practice (Eddisford and Morgan 2018, 154).

As individuals, archaeologists have been known to take part in demonstrations on many issues but rarely do they do so as archaeologists—they generally act in their capacity as citizens or private persons rather than as representatives of an academic or professional expert community. One instance when archaeologists did, however, act as archaeologists was in Britain when English Heritage (the national agency then responsible for archaeology which had control over the site) denied access to various groups to celebrate the Summer solstice at Stonehenge (Bender and Edmonds 1992). In protests at the closing of the site, a number of prominent academic archaeologists were present quoting their authority as experts on prehistoric archaeology to support protesters calls for the site to be made available for those who see the site as sacred (rather than merely as an interesting relic of a distant past or a tourist resource). Negotiations that followed resulted in access being granted on particular terms, and since the restrictions have been less limiting. Here, archaeologists *as* archaeologists made a difference and affected decision-making at very senior levels of government, albeit on matters related directly to issues of archaeology and heritage. As recorded in other sources cited in this book, archaeologists elsewhere work with other disadvantaged and oppressed groups in their capacity as experts on culture to overcome subjugation and denial of rights as human beings.

As McGuire (2008, 16) argues, "archaeology is politics by other means". As such, archaeology can serve a multitude of masters: the state, the international order, particular regimes, interest groups of various kinds (including archaeology itself), those seeking to promote or conduct conflict, those endeavouring to create and maintain peace, and specific communities in specific contexts. What it cannot do—or rather what individual archaeologists cannot do—is to serve antithetical masters at the same time. As those who are inevitably involved in political action—whether we like it or not, agree whether we should or not—we each need to choose where we stand and who beside, both as individual archaeologists and as a disciplinary community. This leads us to consider especially the way archaeology is structured as a community, the topic raised in the next chapter.

References

Aitchison, K. 2009. Archaeology and the Global Financial Crisis. *Antiquity* 83 (319). Available at http://antiquity.ac.uk/projgall/Aitchison319/. Accessed 15 June 2023.

All Party Parliamentary Group on Conservation, Places & People (APPG). 2011. *The Value of Heritage: First Report of the All-Party Parliamentary Group on Conservation, Places & People*. London: APPG/IHBC. Available at https://ihbc.org.uk/appg/resources/CPP-APPF-First-Report%2D%2D-The-Value-of-Heritage-011222.pdf. Accessed 15 June 2023.

Alonso, F.G. 2013. Archaeology and Nationalism: The Development of Archaeology in Catalonia in the Early Twentieth Century. *Complutum* 24 (2): 131–144.

Arnold, B. 1990. The Past as Propaganda: Totalitarian Archaeology in Nazi Germany. *Antiquity* 64 (244): 464–478.

———. 2006. "Arierdämmerung": Race and Archaeology in Nazi Germany. *World Archaeology* 38 (1): 8–31.

———. 2023. *The Perils of a Usable Past: Archaeology's Journey from Culture History to Culture Wars*, CAS Sofia Working Paper Series 14/2023. Sofia: Centre for Advanced Study Sofia. Available at: https://www.ceeol.com/search/article-detail?id=1121865. Accessed 7 June 2023.

Ballbè, E.G., J.O. Puigdoménech, E.S. Olives, and D.W. Steadman. 2007. The Archaeology of the Spanish Civil War: Recovering Memory and Historical Justice. In *Archaeology and Capitalism: From ethics to politics*, ed. Y. Hamilakis and P. Duke, 235–246. London: Routledge.

Bender, B., and M. Edmonds. 1992. Stonehenge: Whose Past? What Past? *Tourism Management*: 355–357.

Bernhardsson, M.T. 2005. *Reclaiming a Plundered Past: Archaeology and Nation Building in Modern Iraq*. Austin: University of Texas press.

Bigambo, R. 2023. Archaeological Heritage Management in Tanzania. *Oxford Research Encyclopedia of Anthropology*. Available at https://oxfordre.com/anthropology/view/10.1093/acrefore/9780190854584.001.0001/acrefore-9780190854584-e-560. Accessed 16 June 2023.

Bower, B.A. 1991. Material Culture in Boston: The Black Experience. In *The Archaeology of Inequality*, ed. R.H. McGuire and R. Paynter, 55–63. Oxford: Blackwell.

British Academy. 2016. *Reflections on Archaeology*. London: British Academy. Available at https://www.thebritishacademy.ac.uk/documents/221/reflections-on-archaeology.pdf. Accessed 15 June 2023.

Carman, J. 1996. *Valuing Ancient Things: Archaeology and Law*. Leicester: Leicester University Press.

———. 1997. Archaeology, Politics and Legislation: The British Experience. In *La Cristalización del Pasado: Genesis y Desarollo del Marco Institucional de la Arqueologia en España*, ed. G. Mora and M. Diaz-Andreu, 125–132. Malaga: University of Malaga Press.

———. 2002. *Archaeology and Heritage: An Introduction*. London: Continuum.

———. 2005. Good Citizens and Sound Economics: The Trajectory of Archaeology in Britain from 'Heritage' to 'Resource'. In *Heritage of Value, Archaeology of Renown: Reshaping Archaeological Assessment and Significance*, ed. C. Mathers, T. Darvill, and B.J. Little, 43–57. Gainesville: University Press of Florida.

———. 2010. Coming Full Circle: Public Archaeology as a Liberal Social Programme, Then and Now. In *Unquiet Pasts: Risk Society, Lived Cultural Heritage, Re-Designing Reflexivity*, Heritage, Culture and Identity, ed. S. Koerner and I. Russell, 151–160. Farnham: Ashgate.

———. 2015. *Archaeological Resource Management: An International Perspective*. Cambridge: Cambridge University Press.

———. 2023. Heritage Values in Comparative Perspective. In *Asia's Heritage Trend: Examining Asia's Present Through Its Past*, ed. J.-I. Kim and M. Zoh. London: Routledge.

Clark, G. 1939. *Archaeology and Society*. 1st ed. London: Methuen.

———. 1947. *Archaeology and Society*, 2nd revised edition. London: Methuen.

Clarke, A. 2017. Heritage Beyond Borders: Australian Approaches to Extra-National Built Heritage. *Archaeologies* 13 (1): 153–174.

Courtney, J.E. 1926. Gertrude Bell (obituary). *The North American Review* 223 (833): 656–663.

Crossland, Z. 2002. Violent Spaces: Conflict Over the Reappearance of Argentina's Disappeared. In *Matériel Culture: The Archaeology of Twentieth Century Conflict*, One World Archaeology 44, ed. J. Schofield, W.G. Johnson, and C.M. Beck, 115–131. London: Routledge.

Cuddy, T.W., and M.P. Leone. 2008. Understanding the Americanization of African Descent Groups Through Archaeology. In *Collaboration in Archaeological Practice: Engaging Descendant Communities*, ed. C. Colwell-Chanthaphonh and T.J. Ferguson, 203–223. Lanham: Altamira Press.

De Cesari, C. 2010. World Heritage and Mosaic Universalism: A View from Palestine. *Journal of Social Archaeology* 10: 299–324.

Dent, J. 2017. Tailors-Made: Heritage Governance Customization in Late Modern Canada. *Archaeologies* 13 (1): 136–152.

Diaz-Andreu, M. 2007. *A World History of Nineteenth Century Archaeology: Nationalism, Colonialism and the Past*. Oxford: Oxford University Press.

Diaz-Andreu, M., and T. Champion, eds. 1996. *Nationalism and Archaeology in Europe*. London: Routledge.

Dietler, M. 1994. "Our Ancestors the Gauls": Archaeology, Ethnic Nationalism, and the Manipulation of Celtic Identity in Modern Europe. *American Anthropologist* 96 (3): 584–605.

Eddisford, D., and C. Morgan. 2018. Single Context Archaeology as Anarchist Praxis. *Journal of Contemporary Archaeology* 5 (2): 117–206.

Eickhoff, M. 2005. German Archaeology and National Socialism. Some Historiographical Remarks. *Archaeological Dialogues* 12 (1): 73–90.

Eze-Uzomaka, P.I. 2000. *Museums, Archaeologists and Indigenous People: Archaeology and the Public in Nigeria*, BAR International Series 904. Oxford: Archaeopress.

Fazioli, K.P. 2012. From First Reich to Third Reich: German Imperialism and Early Medieval Scholarship in the Southeastern Alpine Region (c. 1919–1945). *Archaeologies* 8 (2): 116–144.

Ferguson, L. 1991. Struggling with Pots in South Carolina. In *The Archaeology of Inequality*, ed. R.H. McGuire and R. Paynter, 28–39. Oxford: Blackwell.

Flexner, J.L., and E. Gonzalez-Tennant. 2018. Anarchy and Archaeology. *Journal of Contemporary Archaeology* 5 (2): 213–302.

Funari, P.P., and A.V. de Carvalho. 2011. Inclusion in Public Archaeology in Brazil: Remarks on Collaborative Practices. *Archaeologies* 7 (3): 554–573.

Gänger, S. 2009. Conquering the Past: Post-War Archaeology and Nationalism in the Borderlands of Chile and Peru, c. 1880–1920. *Comparative Studies in Society and History* 51 (4): 691–714.

Gasco, J.L. 2005. Spanish Colonialism and Processes of Social Change in Mesoamerica. In *The Archaeology of Colonial Encounters: Comparative perspectives*, ed. G.J. Stein, 69–108. Santa Fé: School of American Research Press.

Gnecco, C., and P. Ayala, eds. 2011. *Indigenous Peoples and Archaeology in Latin America*. Walnut Creek: Left Coast Press.

Goldsmid, F.J. 1894. Obituary: The Right Honourable Sir Henry Austen Layard, G. C. B. *The Geographical Journal* 4 (4): 370–373.

Goode, J.F. 2007. *Negotiating for the Past: Archaeology, Nationalism, and Diplomacy in the Middle East, 1919–1941*. Austin: University of Texas Press.

Gori, M. 2013. The Stones of Contention: The Role of Archaeological Heritage in Israeli–Palestinian Conflict. *Archaeologies* 9 (1): 213–229.

Green, S., C. Green, and J. Schulderein. 2021a. Archaeology as a Public Good. *Archaeologies: Journal of the World Archaeological Congress* 17 (1) (Archaeology as a Public Good), 1–18.

Green, S., J. Schulderein, C. Green, and N. McDonagh. 2021b. Recreating and Protecting south-east Ireland's Cultural and Environmental Heritage. *Archaeologies: Journal of the World Archaeological Congress* 17 (1) (Archaeology as a Public Good), 19–40.

Greenfield, J. 1989. *The Return of Cultural Treasures*. 1st ed. Cambridge: Cambridge University Press.

Hamilakis, Y., and P. Duke, eds. 2007. *Archaeology and Capitalism: From Ethics to Politics*. London: Routledge.

Harlin, E.-K. 2019. Sámi Archaeology and the Fear of Political Involvement: Finnish Archaeologists' Perspectives on Ethnicity and the Repatriation of Sámi Cultural Heritage. *Archaeologies* 15 (2): 254–284.

Hays-Gilpin, K. 2000. Feminist Scholarship in Archaeology. *The Annals of the American Academy of Political and Social Science* 571. (Feminist Views of the Social Sciences), 89–106.

Hudson, M.J. 2021. *Conjuring Up Prehistory: Landscape and the Archaic in Japanese Nationalism*. Oxford: Archaeopress.

Hutchings, R.M., and J. Dent. 2017. Archaeology and the Late Modern State: Introduction to the Special Issue. *Archaeologies* 13 (1): 1–25.

Hutchings, R.M., and M. La Salle. 2017. Archaeology as State Heritage Crime. *Archaeologies* 13 (1): 66–87.

Johnson, M. 1996. *An Archaeology of Capitalism*. Oxford: Blackwell.

Karl, R. 2020. Authority and Subject (in the Archaeological Discourse in Austria and Germany). *Archaeologies* 16 (1): 137–158.

Karskens, G. 2001. Small Things, Big Pictures: New Perspectives from the Archaeology of Sydney's Rocks Neighbourhood. In *The Archaeology of Urban Landscapes: Explorations in Slumland*, New Directions in Archaeology, ed. A. Mayne and T. Murray, 69–85. Cambridge: Cambridge University Press.

Kiddey, R. 2017. *Homeless Heritage. Collaborative Social Archaeology as Therapeutic Practice*. Oxford: Oxford University Press.

Kiddey, R., and J. Schofield. 2011. Embrace the Margins: Adventures in Archaeology and Homelessness. *Public Archaeology* 10 (1): 4–22.

Kohl, P.L., and C. Fawcett, eds. 1995. *Nationalism, Politics and the Practice of Archaeology*. Cambridge: Cambridge University Press.

Kramer, P.A. 2011. Power and Connection: Imperial Histories of the United States in the World. *The American Historical Review* 116 (5): 1348–1391.

Lafrenz Samuels, K. 2009. Trajectories of Development: International Heritage Management of Archaeology in the Middle East and North Africa. *Archaeologies* 5 (1): 68–91.

Lähdesmäki, T., and V.L.A. Čeginskas. 2022. Conceptualisation of Heritage Diplomacy in Scholarship. *International Journal of Heritage Studies* 28 (5): 635–650.

Laluk, N.C., L.M. Montgomerym, R. Tsosie, C. McCleave, R. Miron, S. Russo Carroll, J. Aguilar, A. Big wolf Thompson, P. Nelson, J. Sunseri, I. Trujillo, G.M. DeAntoni, G. Castro, and T.D. Schneider. 2022. Archaeology and Social Justice in Native America. *American Antiquity* 87 (4): 659–682.

Lightfoot, K.G. 2005. The Archaeology of Colonisation: California in Cross-Cultural Perspective. In *The Archaeology of Colonial Encounters: Comparative Perspectives*, ed. G.J. Stein, 207–236. Santa Fé: School of American Research Press.

Lippert, D. 2006. Archaeology and Integrity: The World Archaeological Congress as a Resource for Indigenous Communities. *Archaeologies* 2 (2): 94–98.

Little, B.J. 2007. *Historical Archaeology: Why the Past Matters*. Walnut Creek CA: Left Coast Press.

Luke, C., and M. Kersel. 2013. *U.S. Cultural Diplomacy and Archaeology: Soft Power, Hard Heritage*. London: Routledge.

Macdonald, S. 2013. *Memorylands: Heritage and Identity in Europe Today*. London: Routledge.

Mayne, A., and T. Murray. 2001. *The Archaeology of Urban Landscapes: Explorations in Slumland*. Cambridge: Cambridge University Press.

McDavid, C. 2004. Towards a More Democratic Archaeology? The Internet and Public Archaeological Practice. In *Public Archaeology*, ed. N. Merriman, 159–187. London: Routledge.

McDonald, K.C., D.W. Morgan, F.J.L. Handley, A.L. Lee, and E. Morley. 2006. The Archaeology of Local Myths and Heritage Tourism: The Case of Cane River's Melrose Plantation. In *A Future for Archaeology: The Past in the Present*, ed. R. Layton, S. Shennan, and P. Stone, 127–142. London: UCL Press.

McGuire, R.H. 2008. *Archaeology as Political Action*. Berkeley: University of California Press.

McGuire, R.H., and R. Paynter, eds. 1991. *The Archaeology of Inequality*. Oxford: Blackwell.

Merriman, N. 2004. Introduction: Diversity and Dissonance in Public Archaeology. In *Public Archaeology*, ed. N. Merriman, 1–18. London: Routledge.

Meskell, L., ed. 1998. *Archaeology Under Fire: Nationalism, Politics and Heritage in the Eastern Mediterranean and Middle East*. London: Routledge.

———. 2002. The Intersections of Identity and Politics in Archaeology. *Annual Review of Anthropology* 31: 279–301.

Meskell, L., and L.S.M. van Damme. 2008. Heritage Ethics and Descendant Communities. In *Collaboration in Archaeological Practice: Engaging Descendant Communities*, ed. C. Colwell-Chthathaphonh and T.J. Ferguson, 131–150. Lanham: Altamira Press.

Mikhailov, D.A. 2013. Altai Nationalism and Archeology. *Anthropology & Archeology of Eurasia* 52 (2): 33–50.

Mrozowski, S.A. 2006. *The Archaeology of Class in Urban America*. Cambridge: Cambridge University Press.

Ndlovu, N. 2016. Fragmented Approach to Governance? Critical Review of the Role Played by Various Government Departments and Agencies in the Administration of Heritage Matters in South Africa. *Archaeologies* 12 (3): 281–303.

Orser, C.E., Jr. 1991. The Continued Pattern of Dominance: Landlord and Tenant on the Postbellum Cotton Plantation. In *The Archaeology of Inequality*, ed. R.H. McGuire and R. Paynter, 40–54. Oxford: Blackwell.

Papoli-Yazdi, L. 2023. Confessions of a Green Notebook: Reading Unpublished Documents About the Oppression of Iranian Archaeology Professors During the 1980s. *Archaeologies*. Available at: https://doi.org/10.1007/s11759-022-09468-9. Accessed 22 June 2023.

Park, H.W., and K. Wee. 2016. The Nationalistic Trend in South Korean Archaeology: Documenting the Development of a Unilinear Evolutionary Trajectory of a Homogeneous Korean Peoples. *Archaeologies* 12 (3): 304–339.

Patton, M. 2007. *Science, Politics and Business in the Work of Sir John Lubbock: A Man of Universal Mind*. Aldershot: Ashgate.

Perring, D., and S. van der Linde. 2009. The Politics and Practice of Archaeology in Conflict. *Conservation and Management of Archaeological Sites* 11 (3-4): 197–213.

Perugini, M. 2008. Anthropologists at War: Ethnographic Intelligence and Counter-Insurgency in Iraq and Afghanistan. *International Political Anthropology Journal* 1 (2): 213–227.

Phillips, C. 1959. Archaeology and the Ordnance Survey. *Antiquity* 33 (131): 195–204.

Pollak, M. 2011. Zur Theorienbildung der archäologischen Denkmalpflege in Österreich. *Österreichische Zeitschrift für Kunst und Denkmalpflege* LXV (3): 227–239.

Pollock, S. 2008. Book Review: Archaeology and Politics. *Diplomatic History* 32 (5): 985–988.

Potter, P.B. 1994. *Public Archaeology in Annapolis: A Critical Approach to History in Maryland's Ancient City*. Washingtom, DC: Smithsonian Institution Press.

Pratap, A. 2009. *Indigenous Archaeology in India: Prospects for an Archaeology of the Subaltern*, BAR International Series 1297. Oxford: Archaeopress.

Price, David H. 2006. Cloak and Trowel. In *Archaeological Ethics*, ed. K.D. Vitelli and C. Colwell-Chanthaphonh, 2nd ed., 116–124. Lanham: AltaMira Press.

Quinn, M. 1994. *The Swastika: Constructing a Symbol*. London: Routledge.

Ralph, J., C. Smith, G. Jackson, I.B. Pamkal, J. Willika, R. Rubio Perez, N. Brown, G. Rankin, A.K. Kanungo, N. Choksi, and The Barunga Community. 2021. Recording Unmarked Graves in a Remote Aboriginal community: The Challenge of Cultural Heritage Driving Sustainable Development. *Archaeologies: Journal of the World Archaeological Congress* 17 (1) (Archaeology as a Public Good), 53–78.

Rankin, L., and G. Gaulton. 2021. Archaeology, Participatory Democracy and Social Justice in Newfoundland and Labrador, Canada. *Archaeologies: Journal of the World Archaeological Congress* 17 (1) (Archaeology as a Public Good), 79–102.

Ratnagar, S. 2004. Archaeology at the Heart of a Political Confrontation: The Case of Ayodhya. *Current Anthropology* 45 (2): 239–259.

Reid, D.M. 2002. *Whose Pharaohs? Archaeology, Museums, and Egyptian National Identity from Napoleon to World War I*. Berkeley: University of California Press.

Rodil, C.J.B. 2020. In Postcolonial Lens: Analysis of Philippine Archaeology's History and Direction. *Archaeologies* 16 (3): 553–594.

Ronayne, M. 2007. The Culture of Caring and its Destruction in the Middle East: Women's Work, Water, War and Archaeology. In *Archaeology and Capitalism: From Ethics to Politics*, ed. Y. Hamilakis and P. Duke, 247–266. London: Routledge.

Rubertone, P.E., ed. 2008. *Archaeologies of Placemaking: Monuments, Memories, and Engagement in Native North America*, One World Archaeology 59. Walnut Creek: West Coats Press.

Rush, L., ed. 2010. *Archaeology, Cultural Property, and the Military*, Heritage Matters. Woodbridge: Boydell Press.

Rypkema, D., C. Cheong, and R. Mason. 2011. *Measuring Economic Impacts of Historic Preservation: A Report to the Advisory Council on Historic Preservation*. Washington, DC: Advisory Council on Historic Preservation. Available at https://www.placeeconomics.com/wp-content/uploads/2016/08/economic-impacts-of-hp_execsummary.pdf. Accessed 15 June 2023.

Sabloff, J.A. 2008. *Archaeology Matters: Action Archaeology in the Modern World*. Walnut Creek: Left Coast Press.

Saunders, R. 2002. Tell the truth: The archaeology of human rights Abuses in Guatemala and the Former Yugoslavia. In *Matériel Culture: The Archaeology of Twentieth Century Conflict*, One World Archaeology 44, ed. J. Schofield, W.G. Johnson, and C.M. Beck, 103–114. London: Routledge.

Schaepe, D.M., B. Angelbeck, D. Snook, and J.R. Welch. 2017. Archaeology as Therapy: Connecting Belongings, Knowledge, Time, Place, and Well-Being. *Current Anthropology* 58 (4): 502–533.

Schlanger, N. 2017. Brexit in Betwixt: Some European Conjectures on Its Predictability and Implications. *The Historic Environment: Policy & Practice*. https://doi.org/10.1080/17567505.2017.1358324. Accessed 8 June 2023.

———. 2018. The Brexit Syndrome: Towards a Hostile Historic Environment? *Antiquity* 92 (366): 1665–1668.

Shepherd, N. 2002. The Politics of Archaeology in Africa. *Annual Review of Anthropology* 31: 189–209.

Silberman, N.A. 1982. *Digging for God and Country: Exploration, Archeology, and the Secret Struggle for the Holy Land, 1799–1917*. New York: Alfred A. Knopf.

Smith, L. 2004. *Archaeological Theory and the Politics of Cultural Heritage*. London: Routledge.

———. 2006. *Uses of Heritage*. London: Routledge.

Smith, A. 2012. A Socially Engaged Archaeology: Spatiality and Governance of Asylum Seekers in Ireland. *Archaeologies* 8 (3): 349–365.

Smith, C. 2017. The Social and Political Sculpting of Archaeology (and Vice Versa). *Pyrenae: Revista de Prehistòria i Antiguitat de la Mediterrània Occidental* 48 (1): 7–44.

Smith, L., and E. Waterton. 2009. *Heritage, Communities and Archaeology*, Duckworth Debates in Archaeology. London: Duckworth.

Smith, C., and H.M. Wobst. 2005. *Indigenous Archaeologies: Decolonising Theory and Practice*, One world Archaeology 47. Abingdon: Routledge.

Spencer-Wood, S.M., and M.C. Trunzo. 2022. Introduction to Archaeologies Special Issue on Intersectionality Theory and Research in Historical Archaeology. *Archaeologies* 18 (1) Special Issue: Intersectionality Theory and Research in Historical Archaeology, 1–44.

Starzmann, M.T. 2013. Occupying the Past: Colonial Rule and Archaeological Practice in Israel/Palestine. *Archaeologies* 9 (3): 546–571.

Starzmann, M.T., S. Pollock, and R. Bernbeck. 2008. Special Issue: Imperial Inspections: Archaeology, War and Violence. *Archaeologies* 4 (3): 353–522.

Steeves, P. 2015. Academia, Archaeology, CRM, and Tribal Historic Preservation. *Archaeologies* 11 (1): 121–140.

Stone, P.G., ed. 2011. *Cultural Heritage, Ethics and the Military*, Heritage Matters. Woodbridge: Boydell Press.

Stone, P.G., and J.F. Bajjaly. 2008. *The Destruction of Cultural Heritage in Iraq*, Heritage Matters. Woodbridge: Boydell Press.

Storozum, M.J., and Y. Li. 2020. Chinese Archaeology Goes Abroad. *Archaeologies* 16 (2): 282–309.

Stottman, M.J., ed. 2011. *Archaeologists as Activists: Can Archaeologists Change the World?* Tuscaloosa: University of Alabama Press.

Thapar, B.K. 1984. India. In *Approaches to the Archaeological Heritage*, ed. H.F. Cleere, 63–72. Cambridge: Cambridge University Press.

Tilley, Christopher. 1989. Excavation as Theatre. *Antiquity* 63. https://doi.org/10.1017/S0003598X00075992. Accessed 23 June 2023.

Todorović, M. 2022. Heritage in and as Diplomacy: A Practice Based Study. *International Journal of Heritage Studies* 28 (7): 849–864.

UNESCO. 1972. *Convention Concerning the Protection of the World Cultural and Natural Heritage*. Paris: UNESCO. Available at https://whc.unesco.org/archive/convention-en.pdf. Accessed 21 June 2023.

Van Helden, D., and R. Witcher, eds. 2019. *Researching the Archaeological Past Through Imagined Narratives: A Necessary Fiction*, Routledge Studies in Archaeology. London: Routledge.
Varouchakis, V. 2017. Indigenous Archaeologies of Crete, 1878–1913. *Public Archaeology* 16 (1): 42–66.
Winter, T. 2015. Heritage Diplomacy. *International Journal of Heritage Studies* 21 (10): 997–1015.
World Archaeological Congress (WAC). 2013. World Congress Newsletter 42, December 2013. http://worldarch.org/wpcontent/uploads/2014/02/enews_42.pdf
Zena, A.G. 2018. Archaeology, Politics and Nationalism in Nineteenth- and Early Twentieth-Century Ethiopia: The Use of Archaeology to Consolidate Monarchical Power. *Azania: Archaeological Research in Africa* 53 (3): 398–416.
Zimmerman, L.J., and J. Welch. 2011. Displaced and Barely Visible: Archaeology and the Material Culture of Homelessness. *Historical Archaeology* 45 (1) (Archaeologies of Engagement, Representation, and Identity), 67–85.
Zoh, M. 2020. *The Impacts of Dictatorship on Heritage Management*. Wilmington D: Vernon Press.

Chapter 7
Archaeology as Social Engagement

It has been a theme of this book from the outset to present archaeologists as members of an international community engaged in an inevitably social activity (see also van Reybrouck and Jacobs 2006). The idea of a solitary archaeologist is something of an oxymoron: as indicated in previous chapters, we work collaboratively in teams when conducting fieldwork, quite often publish collectively, gather at conferences and in other academic settings, form specialist sub-disciplinary groupings, and socialise over food and drink to discuss our experiences. Overall, we are quite gregarious and can be good company. We however are also selective in terms of with whom we choose to mix: there are those we are almost certain to exclude from our company and whose activities we dislike and disdain. As with any community, the creation of archaeology has been a process of simultaneous *inclusion*—of those with whom we deem ourselves to share characteristics—and *exclusion*—of others who we deem not to share those characteristics. The process of selection is of course arbitrary—as is any process of selection for any purpose—because the qualities required for membership are arbitrarily selected from all those available in the universe. We seek those who are like us, are interested in the things we are interested in, do things the way we do them, and who also do not congregate (at least *as archaeologists*) with those we exclude from the discipline of archaeology. (As individuals of course we may congregate with whom we choose: but then we are not acting in the capacity of *archaeologist*).

Elsewhere, I have written that the community of archaeologists "is the anarchist's dream community, a federation of federations in which the smallest decision-making unit is the individual" because "the archaeological community that we… forge is not a single entity but an amalgam of sub-groups" formed by a process in which we "create [disciplinary] sub-groups and sub-sub-groups [of expertise] to which we claim allegiance, ideally to the extent that we [ultimately] stand alone as the sole practitioner of our particular specialism" (Carman 2018, 52). Sinclair, studying patterns of bibliomatic networks in archaeology, described as "bibliographic coupling and co-citation analysis", (2016) suggests a similar idea: that a

J. Carman, *Archaeological Practice as Politics and Ethics*,
https://doi.org/10.1007/978-3-031-69828-6_7

> process of fractal division might exist.... In archaeology, fractal differences can exist because researchers can approach the interpretation of the archaeological record of any period, area or material [in any one of a number of ways]. As archaeological specialists, it is this potential for fractal distinction that separates and unites us as authors within our single discipline and its subfields as much as any distinct set of remains or periods to be explored.

As such, the processes of inclusion and exclusion are applied as much to ourselves and each other as to others. While we all know this, because it is the same process as applied in all other areas of life including other academic disciplines and professions, we do not discuss it or perhaps even acknowledge it. It is nonetheless the manner in which we constitute ourselves, a crucial part of the way archaeology *works* in the world and one key to understanding how and why archaeology is an important set of practices. It is also an element in providing the justification for archaeologists feeling proud of what we do.

Archaeology as a Social Practice

It is unlikely that anyone would challenge the claim that archaeology is a social practice, and indeed to say that it is may be no more than a tautology. A much-used definition of social practice comes from Reckwitz (2002, 249), as "a routinised type of behaviour which consists of several elements, interconnected to one another: forms of bodily activities, forms of mental activities, 'things' and their use, a background knowledge in the form of understanding, know-how, states of emotion and motivational knowledge". Archaeology has all of these: it is only certain types of bodily activity that are identified as 'archaeological', excluding others (see especially Chaps. 2 and 3); it involves thinking in particular kinds of ways but not in others; it involves very particular tools and their use in very specialised ways; it requires a knowledge of how to be an archaeologist (see Chap. 1); it is a skilful activity as demonstrated in a number of ways, both physical and cognitive; it engages us emotionally—we care about what we do and the material we engage with—and are driven to it because of that motivation. The accounts of archaeological practice provided by e.g. Edgeworth and his contributors (Edgeworth 2006) or Cobb et al. (2012), cited elsewhere in this book, all recognise the combination of these elements in our work, as do the aspects of archaeological practice discussed in Chaps. 2, 3, 4, 5, and 6.

- Fieldwork (Chap. 2) involves a well-recognised series of processes that follow each other in logical sequence. It progresses from initial desk-based study, through surveys of various kinds, to excavation, each of which requires a particular form of understanding to be applied. Each of these processes involves a range of tools, from documentary sources, through aerial photographs and scans, to ground-based survey equipment, the laying out of grids with stakes and string, and finally the application of digging equipment, hand-held such as the trowel, hoe, shovel, and mattock; or mechanical such as the digging machine. Survey and digging need to be conducted systematically in order to be effective, and full records made of results and finds in standardised form. Archaeological fieldwork is conducted as team-work in which individuals draw upon each other for assistance and support—and both Hodder (1999) and Edgeworth (2003, 149) have noted the way in which excavators call upon each other to witness their work, eliciting and sharing emotional responses to what is revealed.

- The products of archaeological work (Chap. 3)—reports and other publications, 'grey' literature or otherwise; interpretations of pasts, presents and futures; and commentary on the relations between archaeologists and the material world—are all part of the archaeological canon. Their production is an essential part of the process of doing archaeology—although not by all archaeologists all the time—and require the skilful manipulation of technologies of keyboards and drawing equipment as well as the requisite mental and cognitive processes to create meaningful outputs. This in turn requires the suitable prior understanding and knowledge of what a prospective audience will expect and need. These outputs are often produced collaboratively, and their final publication—in written, spoken, filmic or other form — always elicits a feeling of satisfaction to be shared with colleagues.
- Above all else, archaeologists communicate to and with each other and outsiders (Chap. 4). Whether writing, speaking or acting towards others, archaeologists or not, there is an expectation of what can be expected from specifically archaeological communication. It requires the skilful use of words and images, the manipulation of text- and image-producing technology, and expertise in forms of presentation, whether written, in the form of images, spoken, or performed and enacted. It is founded upon a required knowledge of particular audiences and of what is appropriate for any particular audience and in its particular context. It is driven by desires to achieve certain results, the achievement of which will produce feelings of satisfaction, and may act as a spur to further such communication.
- Archaeology as a public service (Chap. 5)—whether called cultural resource management, archaeological resource management, cultural heritage management, archaeological heritage management, public archaeology (*per* McGimsey 1972) or preventive archaeology—is conducted in much the same manner across the world, involving the same processes (see e.g. Carman 2015) regardless of political, social or economic context. Such work necessarily involves the skills and knowledge applied in fieldwork as well as those to produce outputs to meet the needs of policy or government and public benefit. The motivations are material, in the form of continuing employment and income, but also emotional in the form of the satisfaction of providing a valuable service to those who are the beneficiaries of archaeological provision.
- Archaeology as a form of political engagement (Chap. 6) takes several forms—from reaction to political uses of one's work, to political action of one's own—but is clearly structured in ways that can be identified and if desired replicated by

> others in similar circumstances. As direct action, it may involve physical activity directed outside the archaeological realm, although within an archaeological context; as response it may take the form of the production of written or other outputs, either requiring the use of skills of presentation and communication, drawing upon the tools used in other areas of archaeological work. Any engagement with politics necessarily requires an understanding of the nature of the kind of politics engaged in, its specific context, who are the actors, the outcomes that can be expected and what is the most appropriate means to achieve desired results. The motivation will derive from a desire to alter circumstances for oneself or for others, based upon direct or vicarious experience, and emotional engagement with that experience.

Discussions of social practice in other fields—especially philosophy, psychology, sociology and management studies—are less concerned with defining and identifying forms of social practice than with drawing upon understandings of social practice to change the world for the better (e.g. Welch 2017; Megill 1968). As early as the dawn of the last century, Dewey (1900) was concerned to understand educational practice in such terms in order to change formal education for the better. McCann and Brownsberger (1990) sought to effect change in the way businesses are managed, while Jackson (2005) was concerned with adapting wider social practices to change behaviour towards the environment. Haslanger (2018, 245) relates the understanding of social practice to critical theory and in so doing reformulates the definition of social practice as a resource:

> Social practices are patterns of learned behaviour that enable us (in the primary instances) to coordinate as members of a group in creating, distributing, managing, maintaining, and eliminating a resource (or multiple resources), due to mutual responsiveness to each other's behaviour and the resource(s) in question, as interpreted through shared meanings/cultural schemas.

The constituents of social practices understood in this way are summed up as being "constitutive of social agency, enable coordination around things of value, and are a site for social intervention…. [and are] interdependent in ways that explain their stability, but also indicate sites and opportunities for change"(Haslanger 2018, 247). In other words, established social practices—such as archaeology—provide stability to a community, thus allowing it to maintain itself and develop, but in the course of doing so, also provide the opportunity to change the way the world works.

Beggren and Hodder (2003) discuss aspects of the social practice of archaeology as it has developed historically to identify the origin of the perceived difference between data and interpretation in the field: they argue that lower status accorded 'diggers' (a phenomenon noted in previous chapters and referenced again below) is a cause of this theory/practice divide. Theirs is a concern for the internal workings of archaeology and together with others (e.g. Lucas 2001; Shanks and McGuire 1996) seek to effect change within archaeology itself. For others, archaeology as a social practice that can effect change is directed outwardly (see e.g. Little 2007; Sabloff 2008) and for others yet the aim is to change both archaeological practice and beneficially affect others simultaneously (see Chaps. 3, 4 and 6; also e.g.

McDavid 2004; Hamilakis and Duke 2007; Colwell-Chanthaphonh and Ferguson 2008; Rubertone 2008; Hodder 2010; Ranking and Gaulton 2021; Laluk et al. 2022). Clearly this connects directly with archaeologists' involvement in the political realm and the widespread claim—one also made in his book—that archaeology can be seen as "politics by other means" (McGuire 2008, 16) whether one likes the idea or not. It can of course be counter-claimed—as some do—that archaeology is really nothing to do with politics, that we study the past, preserve what is important from that past, and thereby provide important resources for use by others in the present and future: what use they make of it is not our responsibility, although we should be alert to its misuse. This is a version of the 'passive' response, the claim of 'archaeologist as victim' identified in Chap. 6, but is thereby a retreat from the idea that archaeologists can effect any kind of change in theirs or others' circumstances. This is also a denial of archaeology's capacity as a social practice.

No archaeologist, however, exists alone. The point of this chapter is to emphasise the essentially social and communitarian nature of the discipline. In doing archaeology we perform socially—as part of a wider community of others who also do archaeology. One clear evidence of that is the existence of archaeology as a profession, to be discussed in the next section. This is only one form of how archaeology is organised as a collective body but it is an increasingly important manner of such organisation and represents how many archaeologists see themselves: not merely as members of an academic discipline, but as an economic sector in its own right and one that provides important services to other sectors, both commercial and public.

Archaeology as a Profession

As indicated in Chaps. 1 and 5, as well as elsewhere in this book, those educated and trained as archaeologists may find employment in a number of sectors (and see Chap. 5 on the 'four Cs': Schofield et al. 2011, 100–105). Flatman (2011) lists the available areas of work for the Anglophone world as: cultural resource management; academia; local government; central (national) government; and museums; as well as a range of roles in public archaeology, including school teaching, in lobbying and professional organisations, in non-government heritage preservation agencies, in the media, and providing archaeological expertise for other non-archaeological organisations that require them, both public and commercial, especially as a consultant (Schofield et al. 2011, 105). The positives of working professionally in archaeology are listed as: the rewards of engaging with others equally interested in the past, opportunities for travel, the chance to play a part in protecting and preserving the historic environment, and perhaps influencing public policy. The negatives are: low pay and job insecurity, lack of training or professional development opportunities, few chances to make a name in research, frequently repetitive and dull administrative work, and a lack of clear career structure (Flatman 2011, 181)—points also made in Chap. 1. Advice as to how to prepare for life as a professional archaeologist involves considerable preparation—to explore opportunities for involvement; to read, watch, listen and talk about archaeology; to plan the kind of long-term involvement in the field you may choose; to train appropriately, especially by obtaining relevant formal qualifications most probably at university; to ensure an appropriate level of skill; to volunteer to obtain experience; and finally, and not least, to persevere in seeking opportunities in the face of rejection or criticism (Flatman 2011, 197–200).

The need for perseverance is profiled especially by work conducted on behalf of the EAA among early career archaeologists (ECAs) in all sectors of archaeology, summed up as providing 'a precarious future' for them (Brami et al. 2022). The majority of respondents were based in academia, of whom a fifth also worked at other jobs, the majority in contract and commercial archaeology (working for contractors as defined in Chap. 5: Schofield et al. 2011, 104–105), indicating widespread financial insecurity, with women disproportionally affected (Brami et al. 2022, 231). A majority of those surveyed were pessimistic of finding a secure career, even those with higher levels of qualification, e.g. those holding PhDs or postdoctoral positions (Brami et al. 2022, 233). This was the case both for those most concerned to find a post in academia and those interested in careers in other areas of archaeology. The fact that the survey was carried out for the EAA may have skewed the results towards those more interested in academia as a career, since fewer professional fieldworkers can or do attend the kind of academic conferences organised by organisations such as the EAA or WAC, as confirmed by a survey of attendees of the EAA annual conference in Budapest (EAA 2022): the survey of a majority of attendees established that nearly 70% were based in academic institutions (approximately a third as staff, others as students), and another 22% working in "the heritage sector" (a mixture of permanent and temporary contracts, in public and commercial roles); the remainder were unemployed, retired or working in non-archaeological sectors (EAA 2022, 1). Although over half were provided financial support to attend, only one third were supported by employing institutions, and over one third paid for themselves (EAA 2022, 3).

The EAA survey among ECAs revealed evidence of a lack of opportunity to acquire the necessary skills for a career, especially in academia, particularly limited funding to attend conferences, and few opportunities to gain teaching experience; the authors note especially that "ECRs reported not being able to afford the fees to join professional organizations such as the EAA [so that] those who most need to be members and attend conferences are often the least able to do so" (Brami et al. 2022, 234). There was similarly a lack of assistance in achieving publication or acknowledgement of contributions to research and other projects (Brami et al. 2022, 234) highlighting the need for effective supervision and mentoring for ECAs, as well as much fairer treatment by those more established in the field. Very worrying are the reports of bullying and discrimination, especially but not exclusively affecting women: this ranged from being ignored, through disparaging remarks, to verbal abuse and harassment both physical and sexual; women were also subject to direct discrimination, especially concerning pregnancy (Brami et al. 2022, 235–236). This reflects the essentially 'masculinist' style of archaeology—especially in the field (e.g. Moser 2007) — and its hierarchically managerial nature (Lucas 2001, 8–9), as discussed in Chap. 2. On a more positive note, none attending the EAA meeting in 2022 reported harassment or bullying, which suggests this is probably rare at such events although not unknown (as recent indications from SAA and WAC meetings have shown); and despite the serious problems encountered by archaeologists more generally in the early stages of their academic careers, Brami et al. (2022, 241) nonetheless report that "many ECRs remain hopeful and are driven by love for their respective fields "and that "many point to the collegiality of academic life, the spirit of collaboration, the positives of flexibility, and the excitement at the pursuit of knowledge" (Brami et al. 2022, 242).

As the British Academy (2016, 27–28) notes, "making commercial archaeology more resilient, sustainable, and attractive as a career to new entrants is a key challenge for the twenty-first century" but that the "question of where the responsibility lies for providing training in the skills needed for the modern professional archaeologist continues to be a vexed one" since universities in particular do not usually envisage themselves as providers of training for particular employment. There is a clear disjuncture between the way in which archaeologists are trained—in a formal academic environment with aims of its own—and where the majority will find their employment—in organisations, commercial or otherwise, dedicated to carrying out work related to or deriving from development projects. Although the practices of fieldwork are the same whether in a development-led or pure research context (see Chap. 2), they are very different environments and few university-trained archaeologists are provided with the full range of practical skills and knowledge needed in the development-led world. Flatman (2015, 148) argues that "the rise of the competitive tendering system [in the UK] was also in large part responsible for the more formal separation of the two 'cultures' of 'academic' and 'commercial' archaeology" leading to "the distinction between the 'research' excavation and the 'commercial' excavation, both loaded with formal connotations". This was compounded by the closure of archaeological courses, many of which had been provided as continuing education provision, as well as whole departments of archaeology. The British Academy also notes this trend, emphasising the difference of the UK—where the teaching of archaeology of all kinds tends to be concentrated in a single unit—from elsewhere, where it is divided among other related disciplines; as they put it:

> archaeology departments in the UK normally have expertise across the major periods of the past from early prehistory to the post-medieval period, and across the humanities/science mix of approaches, with British archaeology commonly taught in at least a European and commonly a global context. This inter-disciplinarity is not typical of how the discipline is taught in many countries. In many university systems in Europe, for example, expertise in early prehistory and archaeological science is located with geology and quaternary science, classical archaeology with classics, Egyptology and Assyriology with ancient languages, medieval archaeology with history (British Academy 2016, 17-18).

While treating archaeology as a unity in the UK provides many strengths in terms of disciplinary unity, it also makes them vulnerable as student numbers and research funding reduce: the systems more prevalent in Europe make archaeology less prone to total removal but also serve to divide archaeologists from one another.

The rise of heritage management structures such as preventive archaeology (see Chap. 5) has also had the effect of encouraging the creation of professional groupings of archaeologists. Academic archaeologists have long had their associations where they come together: the Society of Antiquaries of London has its origins in a meeting held in 1707 (https://www.sal.org.uk/about-us/who-we-are/our-history/); its equivalent in Scotland was founded in 1780 (https://www.socantscot.org/); the Deutsches Archäologisches Institut was founded in 1829 (https://www.dainst.org/dai/geschichte); other national and regional organisations also have their origins in the nineteenth century (Diaz-Andreu 2007; Roberts et al. 2020); the International Union of Pre- and Protohistoric Societies (IUPPS) was founded (albeit under a series of different names) at a meeting in Italy in 1866, bringing together various national and period societies (https://www.uispp.org/about/history); the organisation that would be the Society for American Archaeology was founded in 1934 (https://www.saa.org/); WAC was founded—as a break-out organisation from IUPPS—in 1986 (Ucko 1987; Day 2006; Stone 2006); and the EAA was founded in 1994 (https://www.e-a-a.org/). All these institutions focus primarily on organising annual conferences and promote communication and co-operation among archaeologists by supporting publication through book series, journals and newsletters; they also promote best practice by insisting members subscribe to Codes of Practice and Ethics (see Chap. 1).

The first professional association—providing support and accreditation particularly to field archaeologists working outside academia—was the US Society of Professional Archaeologists (SOPA) in 1976 which together with other archaeological organisations in North America re-established itself as the Register of Professional Archaeologists (RPA) in 1998 (https://rpanet.org/about-the-register). Australia followed with the establishment of the Australian Association of Consulting Archaeologists in 1979 (https://www.aacai.com.au/about-aacai/). The UK followed these with the creation of the Institute of Field Archaeologists in 1982, the name changing to the Institute for Archaeologists in 2006, reflecting the wider membership especially of those from academia, and was given formal 'Chartered' status in 2014 hence its formal abbreviation CIfA (https://www.archaeologists.net/about). The Institute of Irish Archaeologists, which is allied to CIfA, was founded in 2001 (https://www.iai.ie/). Although Germany is a country that has no tradition of independent professional organisations, something similar to such an organisation for archaeology was established in 1969 in the Deutsche Gesellschaft für Ur- und Frühgeschichte e.V. (DGUF), which collaborates closely with CIfA and other organisations (see e.g. Scherzler et al. 2019). The alliance of such organisations with CIfA is evidence of the spread—especially promoted by Anglophone archaeologists—of the idea of professional self-organisation, and it has been the subject of discussions of how to reorganise archaeological provision in several countries, including those as widely apart as Korea (KCPIRIA 2014) and Greece (Antoniadou et al. 2018). By contrast with countries in which the state is deemed the proper authority for all public affairs (an issue addressed in Chap. 5), there is a long tradition in the Anglophone world of self-regulating professions, including the law, medicine, accountancy, banking and finance, and commercial administration (see e.g.

Perkin 1989). For archaeologists in Britain to regulate their own developing profession in this way would be appropriate: it is neither unusual nor understood as the appropriation of State authority to private advantage. Elsewhere this is not the case: the role of accreditation in particular often relies upon university qualifications supported by State authority.

Although professional organisation for archaeologists did not emerge until the later twentieth century, the idea of being paid to do archaeological work has a long history. As we have seen in Chaps. 2 and 5, individuals were employed by late medieval Popes and early modern monarchs to seek out ancient sites and material (Carman 2015, 8–9), Everill (2012) charts the work of the 'invisible diggers' the Parkers of Heytesbury in early nineteenth century England, and Mickel (2019) who emphasises the role of expert fieldworkers in the Middle East. These are some of those of which we know: most are anonymous and likely to remain so. The appointment of Augustus Henry Lane Fox Pitt Rivers as the first Inspector of Ancient Monuments in Britain by the legislation passed in 1882 was unsalaried, but it was an official post and could demand limited public financial support: it is the case that Pitt Rivers took full advantage of the latter to fund what were in effect private researches, although the work of excavation was always carried out by workers from his estate rather than his own hands (Chippindale 1983). In the majority of these cases, the credit has gone to the employers rather than those who exercised their skill and knowledge to do the actual work, since it was under the name of the employers that the results were shared. It is perhaps more common in the contemporary environment for development-led work to be published under the name of the employing organisation, although academic research excavations still tend to carry the name of the project lead. As indicated in Chap. 2, fieldworkers have been—and often remain—accorded a lower status than others involved in archaeology.

Professional organisation allows three important linked and parallel processes on behalf of members of the association.

- The first is the generation of codes of practice and standards for work (as mentioned in Chap. 1). These are the formal ethical rules to which members of the association—and by extension all archaeologists, professional or not, members or not—are expected to comply: this is of course a practice of other organisations, more closely related to academia but many of the basic principles are shared (see e.g. Table 1.1). A professional organisation is however much more likely to supplement statements of ethics with guides to actual practice on how to conduct fieldwork, desk-top surveys, post-excavation work, and so on, such as those produced by CIfA (see https://www.archaeologists.net/codes/cifa) or RPA (see https://rpanet.org/code-and-standards). By setting such rules and guidance, the association asserts its own authority and expertise and thus promotes itself as a legitimate arbiter of what constitutes 'good archaeology'.
- The second process is to provide a form of accreditation to its members: membership establishes that members—all of whom are required to comply with the ethical rules and meet at least the minimum standards of performance decreed by

the association—will do 'good archaeology' as defined by the association for clients and employers. It also means that those who are not members cannot be trusted to provide such a high quality of performance, which encourages archaeologists to become members or risk damage to their careers. Associations may either accept qualifications offered by others—such as those provided by academic institutions—or may develop their own as evidence of fitness to join; they may in addition to formal qualifications require evidence of suitable experience; and once established they may also provide schemes of continuing professional development to assist members develop their careers. There may be grades of membership reflecting qualifications and experience, so that perhaps only those deemed capable of directing their own projects will be classed as full members, and others as 'practitioners' or 'associates', so prospective employers can identify those best equipped for the work needed.
- The third is to act as an advocate for the profession, arguing for its practitioners to be accorded higher status and to lead campaigns for both recognition and higher rewards. While any organisation can do this, including those more closely associated to academia, it is specifically the status of the association as a *professional* body that provides the authority to speak. The fact that it sets the standards for performance as an archaeologist, that it accredits those who can be trusted to perform appropriately, and that it represents large numbers of such professionals, all contribute directly to the association's ability to present itself as the main representative body for archaeologists.

But none of these are possible where standards are set by other agencies, where accreditation is the province of other institutions, and where archaeologists—such as where they are automatically servants of the state—cannot negotiate on status or pay. This will inevitably limit the capacity of archaeologists to self-organise to their own advantage.

There is of course more to professionalism than membership in an association, and it is open to all archaeologists—employed or voluntary, academic or in public or commercial service, and at whatever stage in their career—to adopt a fully professional approach to their work. Borrowing from Sullivan (1995), Orser (1997, 252) argues that professionalism combines technical expertise with a wider social purpose so that "Professionalism is an evolving social enterprise that links together the goals and aspirations of individual professionals within a series of complex, interdependent social relationships". Integrity is offered as the key attribute of the professional: integrity "in professional life is not acquired with an academic degree" but where professionals operate as 'good citizens', taking responsible leadership roles in society, the profession will "receive the public's good will, respect and support" (Orser 1997, 253). Here we return to ideas first encountered in Chap. 1 and repeated elsewhere, such as the arguments of e.g. Little (2007) and Sabloff (2008) concerning the contribution archaeology and archaeologists can make to contemporary society.

The Importance of Community

Witmore and Shanks (2013), summarising the results of their interviews with prominent archaeologists, and echoing (albeit a few years in advance) the concerns of this chapter, describe archaeology as "an ecology of practices". In emphasising the diversity that exists within the discipline (evidenced by seven 'scenes' from archaeology's history: Witmore and Shanks 2013, 380–381), they list the relevant elements of this ecology as:

- Institutions, as the structures that hold groups of colleagues together (Witmore and Shanks 2013, 383)
- Affiliates, who are those in other fields with whom we must work to achieve our goals, whilst also accepting the need for compromise (Witmore and Shanks 2013, 384)
- The acquisition and maintenance of competence, through systems of education and the creation of career paths (Witmore and Shanks 2013, 385)
- "Manifesting material pasts" by the transformation of what we encounter into forms of documentation and record (Witmore and Shanks 2013, 390–392; and see also Lucas 2001, 58, cited in Chap. 2)
- "Knowledge design" which determines the forms in which material pasts are made available since otherwise "archaeological knowledge… would not cohere" (Witmore and Shanks 2013, 392; and see Chap. 3 and above on products)
- The means by which the archaeological knowledge design is carried into the future as a form of disciplinary memory (Witmore and Shanks 2013, 393–394)
- Politics, understood as "a sustained commitment to effect positive social change as a component of humanistic and scientific responsibility" (Witmore and Shanks 2013, 394; and see Chap. 6), and
- "The commons" understood as the sharing of what we do and what we make among ourselves and with others.

The glue that holds the diversity of archaeology together, they argue (Witmore and Shanks 2013, 396–398), is its object-orientation and focus on materiality (a point also made by Olsen 2010, cited in Chap. 3), which requires "a freedom of movement and… an intense, diverse inter-disciplinary apparatus [which is] a new practical milieu and become features of a habitat that is completely unique – *archaeology*" (Witmore and Shanks 2013, 398; emphasis in original).

The focus on archaeology as a community of shared interest is highly evident in Witmore and Shanks' (2013) discussion. Drawing upon the influence of Rudwick's (1985) study of the development of early nineteenth century geology, Roberts and Sheppard (2020) emphasise the networks created by and integral to the development of archaeology, and draw especially upon notions of 'thought collectives' (Fleck 1979), actor-network theory (Latour 1987) and geographies of knowledge which locate practices in time and place to inspire their contributors. Similarly, Sinclair (2022) studies archaeology's current collaborative networks using 'science maps', described as "way of representing the nature of academic research based on networks of relationships identifiable from the metadata of academic documents". Such computer-generated maps visualise patterns of authorship, citation and shared language among large numbers (in this case, some 50,000 research outputs). The process reveals archaeology as a "consistently broad and developing subject drawing extensively on methods and approaches from the sciences, social sciences, arts and humanities [and] intrinsically international in practice" (Sinclair 2022). These three examples serve to show a move away from a concern to credit individuals with discoveries and methodological or theoretical advances, and towards the recognition of archaeology as a collective enterprise, which is—in reality—how we experience it.

As outlined above, the creation of formal organisations to provide services to archaeology as a discipline and archaeologists as individuals is an important aspect of the collective nature of the field. Whether a virtual grouping—where membership alone, separate from any particular activity provides the link with others—or a physical coming-together in meetings, lecture halls and social events, such organisations allow archaeologists to claim association with others and assert the status of the individual as part of that collective of specialists. Different organisations serve different purposes. As indicated above, explicitly professional organisations such as CIfA and RPA serve the interests of those employed as archaeologists, and especially those who work in archaeology as a public service (Chap. 5); others, such as the UK-based Prehistoric Society (https://www.prehistoricsociety.org/), Society for Post-Medieval Archaeology (SPMA; https://spma.org.uk/) and or the US-based Society for Historical Archaeology (SHA; https://sha.org/) focus upon those who work in particular periods of the past and are primarily (although not exclusively) providers of conferences and publications. Others again address a wider grouping, usually geographically based—such as the Society for American Archaeology (SAA; https://www.saa.org/), the European Association of Archaeologists (EAA; https://www.e-a-a.org/) and addressing the global archaeological community the IUPPS (https://www.uispp.org/) and WAC (https://worldarch.org/): these host periodic conferences, provide opportunities for publications, may provide financial and other support for archaeologists who are not well placed to meet with international colleagues, and act as advocates on issues of archaeological concern, speaking on behalf of their entire membership. There are also those groupings which are more virtual than real in terms of membership, although managed by a dedicated group of individuals—taking their existence in the form of periodic conferences and the publication of proceedings; examples are the UK-based Theoretical Archaeology Group (TAG; https://www.antiquity.ac.uk/tag) and its international affiliations, and the Contemporary and Historical Archaeology in Theory group (CHAT; McAtackney et al. 2007). Finally, there are those more specialist organisations representing specific multi-period sub-disciplines of archaeology, such as the Association for Environmental Archaeology (AEA; https://envarch.net/), the Historical Metallurgy Society (HMS; https://historicalmetallurgy.org/) or those combining a focus on period with that on material, such as the Medieval Pottery Research Group (MPRG; https://medievalpottery.org.uk/). All of these organisations combine long-term association by individuals with short-term memberships by others to take advantage of particular opportunities, such as conference attendance.

As indicated in Chap. 4, there is more to the conference than to present papers and listen to those presented: as Witmore and Shanks (2013, 395) put it, they also provide "atmospheres for swapping gossip in hallways, across tables, or along bars [and as might be added, over stalls in book rooms and in toilets]. Through such rituals [archaeological] organisations... seek to strengthen one's sense of belonging to the archaeological community". The conference is also one of the locales in which we learn to become part of that archaeological community: in the same way as we are introduced to the codes of behaviour of archaeology in fieldwork training (Chap. 2), the conference is the place where we learn how to meet each other, how to engage with others whose work or context for work is different from our own, and how to behave, dress, speak and listen in the academic environment. While fieldwork (see Chap. 2) is highly organised with a clear hierarchy of authority, in the conference all are deemed to be equal: it is not uncommon for participants to be housed in the same place (either a hotel or student rooms) and to share meals in the same dining hall (especially breakfast). Travel to the conference venue may be by organised coach, or it may be reached by a shared walking route, which will also bring participants together, and any organised field trips or guided institutional visits (e.g. to local museums, sites or academic institutions) will enhance the sense of togetherness. The result is a shared experience of the event, from which non-participants are excluded. The common practice in some parts of the world—offered especially by international organisations—to provide certificates of attendance adds enhanced status both to the event and to individuals: it confirms the official and institutionally-accredited status of the event and serves to emphasise its professional functions, as part of an individual's developing *curriculum vitae*. There is also the practice among some—especially younger scholars—to include membership of a conference organising institution (e.g. the EAA, SAA, WAC) as part of their CV, offering it as if proof of professional accreditation, although such organisations are not 'professional' in the sense of CIfA or RPA which do provide accreditation for their members: certificates of attendance can then support such use of the institutional name.

Drawing upon similar experience, Deardoff (2015, 318) notes that "we should not neglect to consider that the purpose of the [academic] conference is much larger than merely research dissemination, and that the constituencies impacted by new structures are much broader than the tenured faculty at research institutions". This is an important point in considering the rise of the virtual and hybrid conference, occasioned by the emergence of reliable conferencing technology and especially following the global Covid-19 crisis. As noted in Chap. 4, those who cannot or choose not to attend in person are denied all but participation in formal exchanges: you cannot share a drink or a meal very effectively through a computer screen, participation in a free-flowing conversation with those present with each other in person is more difficult, and opportunities for chance encounters non-existent. While it is true—as Deardoff (2015, 316) notes—that "face-to-face meetings are not essential for network construction or maintenance" because it is perfectly possible to conduct research and publish jointly with those one has not met except via computer, nonetheless "in-person meetings at conferences may create opportunities that may not be replicable in a virtual world": the 'Chat' function of systems allowing live conversation online is hardly conducive to the informal talk that is the hallmark of real chatting.

As indicated earlier and in Chap. 2, fieldwork is inevitably a shared enterprise: it is conducted in teams (or crews as they are frequently called in North America) and the outcomes are the product of that collective endeavour (Chap. 3). Elsewhere (Carman 2006, 97) I have emphasised the similarity between archaeologists working on site and teams of building workers or the crews of a ship at sea: these too operate in an environment set apart from the rest of the world; they will not only work together but frequently also eat and relax together (Carman 2006, 100–101). This separation from the rest of the world—whether on an enclosed urban site or out in the 'wild'—is integral to building a sense of common purpose by creating the image of the site as a place where "something unusual, special, and indeed interesting is taking place" (Carman 2006, 97). On site, the focus is inwards, on the site and what takes place there: it is rarely to look about and out of the site, because to do the latter might involve encounters with others not part of the activity. In such contexts, 'the public' are outsiders—they may also look into the site, but always from the outside; we are insiders who can see it for what it truly is. As Chap. 2 emphasised, fieldwork and its associated activities are as constitutive of the archaeological community as any other activity we perform.

The outputs of archaeological fieldwork (Chaps. 3 and 4) also serve to form and consolidate the community that is archaeology. Joint publication—especially of site reports and academic outputs, such as books and journal articles—is common in the field; and single-authored outputs tend to contain extensive acknowledgements to those who played a part in developing the work. Sinclair (2022) lists the manners in which archaeological community by publication is established and maintained; in the case of identifying collaborative networks, they are also the indicators of such networks:

- Direct citation—from one document back to a source document
- Ego-Alter Citation, which distinguishes between the references that an author uses in the course of their research, and the way in which their own documents are made reference to by others
- Bibliographic coupling, which links together two or more citing documents that share a number of cited references in common
- Co-citation, which links together two or more cited documents that are referenced together in another document
- Co-authorship, which links people who collaborate as research partners
- Co-occurrence relationships, which identify academic elements that occur together, such as terms used for describing the research of a discipline (Sinclair 2022).

The maps of disciplinary connections thus drawn identify common areas of work and their differences from others: comparison between maps made for different periods shows how the field has developed. It also reveals "the breadth of [archaeology's] sources and range of its terms, and their development through time, demonstrating that archaeology is a discipline that openly draws in opportunities to expand its research methods and direction through engagement with practices in other disciplines" (Sinclair 2022). The work also suggests that this breadth and diversity may require us to consider our use of language to describe our work: perhaps to create a common lexicography in order that we communicate within the field most effectively (and see also Venclová 2007). The work also reveals "significant variations in male/female representation by research specialty" (Sinclair 2022) leading to a need to study "the impact of specific career trajectories and the influence of role models or mentors within particular areas of research" (Sinclair 2022; and see also Heath-Stout 2020).

The work of Sinclair (2022) indicates what we already know about our discipline: that it is highly diverse, and open to draw upon ideas from other disciplines. The diversity is evidenced by and reflected in the many sub-disciplinary groupings that exist in archaeology as a global enterprise: these in turn are reflected in the research groups and organisations discussed above, drawn along geographic, period, methodological or material concerns or combinations of them. Archaeologists are always developing new branches of the field, and while some can be incorporated straight away, sometimes as a sub-discipline of an established expertise, others may need to struggle to reach acceptance. As such new fields emerge, they can develop in one of two ways: as in the case of Conflict Archaeology (Carman 2018, 49–51), by separation from other branches until given due recognition as legitimate; or as in the case of public archaeology (Carman 2018, 47–49), by inserting discussions into others' conferences and conference sessions, sometimes clandestinely by deliberate false labelling, until accepted. In both cases, it takes a measure of courage and determination to obtain acceptance, and also some time: it is unclear whether some forms of Conflict Archaeology are yet accepted by some other branches of the discipline, especially because of the connections with the metal-detecting community; and while there is much more acceptance of public archaeology and heritage as legitimate areas for debate, there remain those, especially among senior staff of some national state agencies, who are unhappy at academic critique (even though not all of it is critical in the pejorative sense).

Whether figuratively huddling together against the intellectual assaults of those who disagree with us and devalue our efforts, struggling to understand and interpret the material residues of past lives or past communities in sometimes inclement conditions, endeavouring to obtain recognition for our work through publication and outreach, or simply communicating to sympathetic co-workers what we do, archaeology is always a social and collective enterprise. The idea of 'the archaeological community'—a phrase widely used—is no invention: archaeologists do form a community. It is one of the few disciplines where widely divergent practitioners in its sub-groupings can talk to each other meaningfully about their work, and be mutually understood. Sinclair (2022) may be right that we need a common set of terms so we can each fully comprehend the relevance of others' work to our own, but experience shows that we can talk amongst ourselves and be understood quite readily. We need to recognise, however, that however open we may be to draw upon the knowledge and ideas of others, not all are welcome among us.

Archaeological Communities—The Included and Excluded

There are perhaps four loose categories of persons with whom archaeologies engage. Put broadly we can distinguish three such categories from the fourth, which is we ourselves: the latter is Us, the remainder a Them. These fall into three categories: those we deem to be actively hostile to our work, and to whom we are hostile in return, 'enemies'; those who we deem as no threat but are also quite separate from us, 'neutrals'; and those who we deem to be of assistance to us in achieving our objectives, 'allies'. Having thus divided the world beyond us so neatly, we then need to recognise that nothing is quite so simple: in fact anyone falling into any of these broad categories can move into another by the simple expedient of a change of perspective on our part or a change of behaviour on theirs. Table 7.1 sets out this initial categorisation.

Among the threats to archaeology, metal detectorists, looters and treasure hunters are frequently treated as a single category but they are in fact quite different. It is true that each may damage or destroy valuable archaeological resources, and any of them may appropriate what they discover for their own advantage, which can be financial. The word looter indicates a person whose activities result in the loss of sites and material; the treasure hunter seeks to take valuable (that is, financially valuable) material and objects for their own advantage; and the metal detectorist uses a particular device—a metal detector—to seek objects of metal that are of interest to them, which can be quite mundane objects or those that are archaeology significant or financially valuable. Any one may also be one of the others—a looter or treasure hunter may use a metal detector; a metal detectorist may loot archaeological sites for treasure — but the distinctions concern the consequences (looter), the motive (treasure hunter) and the method (metal detectorist). Each category also has its own sub-categories: among the looters can be included the *tombaroli* of Italy (van Velzen 1996), also merged with the so-called *clandestini* by Renfrew (2000, 17) although they not be always the same. Van Velzen's (1996) discussion of the *tombaroli* is enlightening: it reveals them as a form of popular hero in their own eyes

Table 7.1 Categories of person as perceived by archaeologists

Enemies	Neutrals	Allies	Ourselves
Metal detectorists	The public	Indigenous communities	Academic archaeologists
Looters		Descendant communities	Professional archaeologists
Treasure hunters		Local communities	Students
Collectors			Volunteers
Developers			Amateurs
Military			

and that of their compatriots, poor people who become rich and thereby worthy by their own efforts, who rescue the Etruscan past from the state, publish their life stories and appear on television. They present themselves as having "a bond with their past…. [and] that financial gain is not the main motivation behind their work, but that they are driven by an affinity with their forebears which develops during their nocturnal excavations" (van Velzen 1996, 124).

Others categorised as looters include those charged with destroying sites and monuments for political reasons, such as the Taliban destruction of the Bamiyan Buddhas in Afghanistan or destruction by ISIS of sites in various parts of the Middle East and Africa, designed to assert ascendancy of a particular understanding of Islam (Harmanşah 2015; Campion 2017), or as part of wider programmes of cultural genocide in the Balkans (Walasek 2020) or the Uyghurs of China (Ruser et al. 2020), among others. ISIS in particular have been charged with using the looting of sites as a source primarily of funding (e.g. Terrill 2017), a charge that while probably true also serves to reduce any moral claim they may have to their activities: if the destruction of parts of the World Heritage Site at Palmyra was intended as a challenge to Western dominance in the region, their reduction as a result of the charge to mere profiteers from such damage renders them no more than conventional criminals rather than a dedicated Islamist resistance movement. By contrast, the looting of Iraq's National Museum following the US-led invasion represented two things: an outpouring of anger at the defeated regime and a desire to seize objects of value for sale in the market (Brodie 2006). It is clear from this that looting is not a simple matter to understand: it is driven by a range of motives, some complementary, others contradictory, and operating independently or simultaneously. Nonetheless, there is an archaeological consensus that looting and the damage to sites is inevitably a wrong.

Before the metal detector emerged as a major threat to the archaeological record, much of the focus of concern was levelled at treasure hunters, both on land and especially at sea (e.g. Cockrell 1980; Gould 2000; Perez-Alvaro 2019, 214–228). As Gould (2000, 27) explains, "recovery of antiquities by treasure hunters for purposes of commercial gain, either through direct sale of artifacts or indirectly through media exploitation, is arguably the single most destructive influence on the underwater archaeological record". Perez-Alvaro (2019, 218–220) emphasises particularly the legal issues, and especially the weaknesses of regulation which rely on a combination of national laws and international agreement, which leave gaps in coverage. She emphasises also the issue of the interference with wrecks, which are also frequently grave sites (Perez-Alvaro 2019, 135–158). A paper presented to the seventh Fields of Conflict conference (Czakó 2012) brought the topic to the fore: artefacts recovered from ships of the sunken Austro-Hungarian World War I fleet were presented as trophies, and plans to raise one for use as a floating restaurant offered as if ethically unproblematic; while the organisers—close connections with the Hungarian military—saw no problems, international archaeologists present had major concerns. Perez-Alvaro also discusses problematic cases with ethical as well as legal aspects: notably the use of lead ingots from Roman wrecks as shielding for scientific experiments because of their low atomic emissions (Perez-Alvaro 2019,

68–79). Whether all such appropriation of underwater material counts as 'treasure hunting' becomes questionable but, as with looting, there is archaeological consensus against treasure hunting.

While the use of metal detectors is banned outright in some territories, and in most other territories disallowed in certain contexts but allowed in others, the metal detectorist can be seen as either an enemy of archaeology—a member of the ranks of the looter and the treasure hunter—or as a potential ally due to the usefulness of the tool as surveyed by Connor and Scott (1998). Where there is an outright ban, its use is denied equally to the archaeologist and the amateur, but the majority of limits are placed on what the machine can be used for: usually the retrieval of categories of protected material. The British Portable Antiquities Scheme (Bland et al. 2017) which encourages metal detectorists—and others who retrieve objects of archaeological interest—to have them officially recorded is having some influence in territories more suspicious of detecting; Belgium has sought to prevent its use except by consent of the authorities but is now considering lifting such restrictions. Elsewhere (Carman 2004, 49–50), I have pointed out that while metal detectorists may not do good archaeology they may not consider that they do real harm: unlike us they do not occupy sites for any length of time, and they extract only what is of interest to them, rather than destroying all remains, and clearing large amounts of soil. Thomas (2012, 75–76) indicates that a majority of British metal detectorists are interested in working with—rather than against—archaeologists and there are attempts at self-regulation to limit the activities of so-called nighthawks who operate clandestinely to damage sites. She goes on (Thomas 2012, 77) to suggest that there is a shift in archaeological responses to metal detecting which is related to the rise in concerns for public engagement (Chap. 3; Skeates et al. 2012b; Moshenska 2017).

That unrestricted development can—and will—damage and destroy potentially important archaeological remains is well recognised, hence the provision of legislation across the globe to offer protection and place limitations on such activity (as discussed in Chap. 5). Preventive archaeology in its various forms—as also discussed in Chap. 5—serves to ally archaeology to development by ensuring that prospective development will not be halted by the unforeseen discovery of significant remains during building work. Nonetheless, antipathy between those who wish development to proceed and archaeologists remains: Cooper (2008) outlines some of the ways in which that antipathy is mobilised against archaeologists, which he calls 'rhetorical destruction'. He lists the ways in which various discursive devices are applied to challenge the authority and expertise of those working as curators of the archaeological resource (a role discussed in Chap. 5):

- By discrediting the resource under consideration (Cooper 2008, 21–22)
- Discrediting cultural resource philosophy and practice (Cooper 2008, 22–24)
- Discrediting organisations involved in preservation (Cooper 2008, 24–26)
- Appeals to local sympathies and connections of various kinds which are presented as overriding heritage preservation values (Cooper 2008, 26–28).

The last set of these rhetorical tricks place a concern for public engagement over the past in opposition to an interest in protecting local pasts. The others play on what may be longstanding distrusts of officialdom and the imposition of controls by a possibly distant government. They can be presented, if done cleverly, as a disinterested response to outside interference in local matters: and as Cooper demonstrates with an example, can be successful (Cooper 2008, 17–18).

The ambiguous relation of archaeologists to the military has been discussed in Chap. 6 (and see Stone and Bajjaly 2008; Rush 2010; Stone 2011a). While some archaeologists are prepared to work with the military in an effort to protect the archaeological resource during military operations, others find the very idea offensive (as described by Stone 2011b, 4–9). Others yet work actively with military organisations in other endeavours—such as using archaeology as a form of therapy (e.g. Brizi et al. 2023) or in recognising the military as organisations and people who are directly concerned for their own heritages which are also those of a wider community (Brown 2011). Nonetheless, the record of military activity in relation to archaeological resources is frequently one of damage and destruction: the use of munitions causes widespread damage to sites and monuments. Sites, because of location or form, are frequently used for military purposes, whether it is as defensive or observation positions (e.g. the Butte de Warlencourt in World War I: Saunders 2004, 10–11) or as depots for personnel or *matériel*. Long-established military establishments are also repositories of archaeology and subject to efforts at preservation and record (e.g. at Fort Hood, Texas: Barratt et al. 2006). As also noted in Chap. 6, archaeologists of earlier eras have also been connected to soldiering one way or another and several prominent innovators—Pitt-Rivers, Wheeler and others—have served as military officers. The military, therefore, is one of those institutions which although in some ways antithetical to archaeology can also be considered in other contexts as either a neutral or an ally. The decision as to how to consider them is primarily an ethical one (as noted in Stone's 2011a title): as always with such choices, it is not easily made and there is no absolute right or wrong, or indeed generally agreed, answer.

The public—a generic and in some ways purely residual category (*per* McManamon 1991)—is classed here as neutral regarding archaeology because there is, as outlined in Chap. 3, a perceived need among us to educate its members (e.g. Cleere 1984; McGimsey 1984) even if not explicitly perceived as 'deficient' (as outlined by Merriman 2004). While 'the public' as a category defined in terms of an agglomeration of individuals may not understand or appreciate the work of archaeologists, it is capable of being persuaded as to our value by suitable processes of outreach (discussed in Chap. 4). On the other hand, 'the public' as a realm of activity to be distinguished from the private (Chap. 5; and see Merriman 2004) is fully cognisant of our value as makers and preservers of our collective pasts. It is also worth emphasising here how recent has been the concern to involve publics (in the sense of communities of other human beings) in what we do; for much of the history of our field the public as other people was considered something of an irrelevance. However, once engaged by us agglomerations of other humans become active supporters of our cause and indeed collaborators in our activities. These are not a generic public but rather identifiable communities—local or descendant (Marshall 2002), Indigenous (Rubertone 2008) or otherwise (Colwell-Chanthaphonh and Ferguson 2008)—with whom we engage in various ways (Smith and Waterton 2009) for our benefit and theirs (and as discussed in Chaps. 3, 4 and 6).

Finally, there is the definition of ourselves. As indicated in Chap. 1, it depends on your own context as to who may be considered 'one of us'. In those countries with a long tradition of volunteer and amateur activity, and despite the strictures of the Valletta Convention (Council of Europe 1992) that only allow those suitably qualified to excavate, it is possible for anyone to declare themselves to be an archaeologist regardless of a lack of formal training. In others, where amateur archaeology is unheard-of, to make such a declaration will be open to direct challenge and even possibly to sanction. Across the globe, those employed as archaeologists in academia or cultural resource management will be accepted as archaeologists: the inception of professional organisations providing accreditation based on demonstrable experience and skill even in the absence of formal university education serves to enhance the drive towards the recognition of the field as one for professionals and thereby to reduce opportunities for volunteer involvement. Membership of a professional body—and it is noticeable how primarily academic organisations such as the SAA, SHA, EAA, IUPPS and WAC increasingly refer to themselves as associations of professionals—is increasingly used as evidence of qualification to be considered an archaeologist. Our ranks close around us.

The aim of this section has been to emphasise how archaeologists as a community define not only ourselves but also others to suit our own interests (a point I have made elsewhere: Carman 2015, 178–180). All the categories above—not merely the arbitrary relational categories of 'enemy', 'neutral' and 'ally' but also some of the more specific descriptions of 'metal detectorist', treasure hunter', 'looter', 'public' and 'community'—are all quite arbitrary: nobody is automatically hostile to archaeology, and nobody automatically friendly towards us, nor neutral, nor indeed necessarily 'one of us'. It depends very much on specific contexts and on our attitudes towards others and their activities. Perkin (1989) writes of how the middle-class commentators on late nineteenth century Britain effectively wrote themselves out of the story they created, defining the various classes in doing so: we tend to do the same in failing so often to recognise the role we play in defining others in relation to ourselves. As in so much of what we do and how we do it, this is an ethical choice we make about inclusion within and exclusion from among us, and needs to be acknowledged as such.

Archaeology in Society

Implicit in the previous sections—and indeed in large parts of the book so far—archaeology has been treated as something separable from the wider world of which we are nonetheless a part. As Chaps. 3, 4, 5, and 6 have demonstrated and discussed, in various ways archaeology is always looking beyond itself either to influence or to serve: whether as a producer of outcomes or of products (Chaps. 3 and 4), as a branch of the public realm (Chap. 5), or as an engagement with politics (Chap. 6), our work engages with those outside our own field, even to the extent (as discussed above) of determining those practices and persons who can be treated as part of our own endeavours. However we seek to define ourselves (and see Chap. 1), we do so in the context of those who surround us and with whom we engage even as we seek to distance ourselves from them: as McGimsey (1972, 5) pointed out "There is no such thing as a 'private' archaeology" and the idea of archaeology as a *res publica* (a public thing) is implicit in all that we do: to deny the public nature of archaeology is to place ourselves in the position of those who explore the past entirely for their own gratification and benefit—putting us among the 'looters' and 'treasure hunters' (such as Czakó 2012) we otherwise seek to condemn.

Schofield (2010) lists some of the ways in which archaeology benefits contemporary society, all of which have appeared elsewhere in this book: as public and community outreach and collaboration; as part of a wider education; in helping communities understand themselves and their identities; in providing models for how to live better and more sustainability; and providing deeper understanding of how humans interact with a material world. In arguing for an archaeology of the contemporary world (and see Smith 2017), we are told that "Archaeology presents an important opportunity…to benefit society now and in the future [and] opportunity and responsibility to reflect on the role and relevance of archaeology, not necessarily in response to detractors and hecklers, but with a view to renegotiating the place of archaeology in society" (Schofield 2010, 327). Reflecting upon the effects upon archaeology of the creation of WAC in 1986, Zimmerman (2006, 87) suggests that we have "sought to liberate archaeology from its practice of scientific colonialism…. [and that] forces of liberation in archaeology reflect ongoing national and international trends away from colonialism, toward self-determination and a local control of most aspects of peoples' lives". In similar vein, Thomas et al. (2014, 1, emphasis in original) present "archaeology and heritage projects as, in effect, *conversations*" among a collection of different types of persons—participants in archaeological work, volunteers, practitioners, professionals and academics. Skeates et al. (2012a, 8) echo such sentiments by a concern that "we might all contribute to defining archaeology's place in the world".

As I have commented elsewhere (Carman 2018, 45–46), there is a periodic tendency among archaeologists to examine ourselves—a trend we have noticed in several previous chapters especially as self-critical perspectives on aspects of practice. This is of course a positive aspect of archaeological practice, especially as it concerns our relations with others, and it is as relations with others that such self-examination is most frequently encountered. Nillson Stutz (2018, 54–55) argues that we are "ideally situated to lead debates about cultural heritage, identity politics, postcolonialism, migration, the Anthropocene and its challenges, and cultural change – all central themes for some of the most urgent intellectual and political debates in contemporary" life but it is also suggested that "archaeology is held back by its own 'brand'". The answer that is offered is to "change the game by driving a collaborative transdisciplinary archaeology that truly integrates multiple perspectives to understand the past, and that insists on problematising it – and our contemporary world – not on simplifying it" (Nilsson Stutz 2018, 55). In similar vein,

Prescott (2013) argued for closer engagement between those in the humanities and those in the sciences. Mutual comprehension is the key here, since "Communication is based on knowledge, and knowledge is based on the insights training engenders in theory, assumptions, methods, applications, validity and relevance.... The science competences that archaeologists will have to possess to be qualified partners in knowledge production in" the future is a priority in archaeological education (Prescott 2013, 43).

One of the concepts that archaeologists across the globe have taken to heart is that of 'sustainability'. As Favreau and Patalano (2017) indicate in their collection from conference presentations, this generally takes two forms: a concern for working with and creating sustainable communities in the investigation of their pasts; and developing techniques and methods that will make archaeology itself sustainable and assist in the preservation of material from the past. This of course reflects concern introduced in Chap. 1, especially from the suggestions of Little (2007) and Sabloff (2008) among others concerning the values of archaeology to contemporary society. Archaeology is integrally incorporated in a social context and that context is at once global and local: we are a society of our own making which is part of the wider world and to which we make important contributions. Any emphasis on our role as creators of pasts alone is to sell us short: as Chap. 3 indicated, we also make presents and futures. Any emphasis on technical skills—whether fieldwork-based or interpretative—to the exclusion of social, political and economic importance is to sell us equally short. To deny our role in public service—which goes beyond official functions in terms of archaeological resource management—is to sell us short again. And to deny that archaeology is anything but a political enterprise—and always has been—is to minimise our place in the world. We are much more important than that: the next and final chapter will draw together the underlying themes of the book to emphasise this.

References

Antoniadou, S., G. Vavouranakis, I. Poulios, and P. Raozaiou, eds. 2018. *Culture and Perspective at Times of Crisis: State Structures, Private Initiative and the Public Nature of Heritage*. Oxford: Oxbow.

Barratt, G., L. Dingwall, V. Gaffney, S. Fitch, C. Huckerby, and T. Maguire. 2006. In *Heritage Management at Fort Hood, Texas: Experiments in Historic Landscape Characterisation*, ed. L. Dingwall and V. Gaffney. Oxford: Archaeopress.

Berggren, Å., and I. Hodder. 2003. Social Practice, Method, and Some Problems of Field Archaeology. *American Antiquity* 68 (3): 421–434.

Bland, R., M. Lewis, D. Pett, R. Richardson, K. Robbins, and R. Webley. 2017. The Tresure Act and Portable Antiquities Scheme in England and Wales. In *Key Concepts in Public Archaeology*, ed. G. Moshenska, 107–121. London: UCL Press.

Brami, M., S. Emra, A. Muller, B. Preda-Bălănică, B. Irvine, B. Milić, A. Malagó, K. Meheux, and M. Fernández-Götz. 2022. A Precarious Future: Reflections from a Survey of Early Career Researchers in Archaeology. *European Journal of Archaeology* 26 (2): 226–250.

British Academy. 2016. *Reflections on Archaeology.* London: British Academy. Available at: https://www.thebritishacademy.ac.uk/publications/reflections-archaeology/. Accessed 4 July 2023.

Brizi, A., A. Rabinovich, and C. Lewis. 2023. Psychological Outcomes of Local Heritage Engagement: Participation in Community Archeological Excavations Increases Well-Being, Self-Efficacy, and Perceived Community Support. *Journal of Applied Social Psychology* 1–12: 850–861. https://doi.org/10.1111/jasp.12972.

Brodie, N. 2006. The Plunder of Iraq's Archaeological Heritage, 1991–2005, and the London Antiquities Trade. In *Archaeology, Cultural Heritage, and the Antiquities Trade*, Critical Heritage Studies, ed. N. Brodie, M.M. Kersel, C. Luke, and K.W. Tubb. Gainesville: University Press of Florida.

Brown, M. 2011. Whose Heritage? Archaeology, Heritage and the Military. In *Cultural Heritage, Ethics and the Military*, Heritage Matters, ed. P.G. Stone, 129–138. Woodbridge: Boydell Press.

Campion, K. 2017. Blast Through the Past: Terrorist Attacks on Art and Antiquities as a Reconquest of the Modern Jihadi Identity. *Perspectives on Terrorism* 11 (1): 26–39.

Carman, J. 2004. Excavating Excavation: A Contribution to the Social Archaeology of Archaeology. In *Digging in the Dirt: Excavation in a New Millennium*, British Archaeological Reports International Series 1256, ed. G. Carver, 45–51. Oxford: BAR.

———. 2006. Digging the Dirt: Excavation as a Social Practice. In *Ethnographies of Archaeological Practice: Cultural Encounters, Material Transformations*, Worlds of Archaeology, ed. M. Edgeworth, 95–102. Walnut Creek: Left Coast Press.

———. 2015. *Archaeological Resource Management: An International Perspective*. Cambridge: Cambridge University Press.

———. 2018. 'Herding Cats': Building Archaeological Communities. *Journal of Community Archaeology & Heritage* 5 (1): 45–54.

Chippindale, Christopher. 1983. The Making of the First Ancient Monuments Act, 1882, and Its Administration under General Pitt Rivers. *Journal of the British Archaeological Association* 136: 1–55.

Cleere, H.F. 1984. World Cultural Resource Management: Problems and Perspectives. In *Approaches to the Archaeological Heritage*, ed. H.F. Cleere, 125–131. Cambridge: Cambridge University Press.

Cobb, H., T.J. Harris, C. Jones, and P. Richardson, eds. 2012. *Reconsidering Archaeological Fieldwork: Exploring On-site Relationships Between Theory and Practice*. New York: Springer.

Cockrell, W.A. 1980. The Trouble with Treasure. A Preservationist View of the Controversy. *American Antiquity* 45 (2): 333–339.

Colwell-Chanthaphonh, C., and T.J. Ferguson, eds. 2008. *Collaboration in Archaeological Practice: Engaging Descendant Communities*. Lanham: Altamira Press.

Connor, M., and D.D. Scott. 1998. Metal Detector Use in Archaeology: An Introduction. *Historical Archaeology* 32 (4): 76–85.

Cooper, M.A. 2008. This Is Not a Monument: Rhetorical Destruction and the Social Context of Cultural Resource Management. *Public Archaeology* 7 (1): 17–30.

Council of Europe. 1992. *European Convention on the Protection of the Archaeological Heritage (Revised) (ETS No. 143)*. Available at: https://rm.coe.int/168007bd25. Accessed 28 Nov 2022.

Czakó, L. 2012. *The Research of the Monarchy's Shipwrecks*. Paper presented to the 7th Fields of Conflict conference, Budapest, Hungary, 18th to 20th October 2012.

Day, M. 2006. Peter Ucko and the World Archaeological Congress (WAC). In *A Future for Archaeology: The Past in the Present*, ed. R. Layton, S. Shennan, and P. Stone, 41–46. London: UCL Press.

Deardorff, M.D. 2015. Reconsidering the Scholarly Conference for the Contemporary Academic. *PS: Political Science and Politics* 48 (2): 315–318.

Dewey, J. 1900. Psychology and Social Practice. *Science N.S.* 11 (270): 321–333.

Diaz-Andreu, M. 2007. *A World History of Nineteenth Century Archaeology: Nationalism, Colonialism and the Past*. Oxford: OUP.

Edgeworth, M. 2003. Acts of Discovery: An Ethnography of Archaeological Practice. In *BAR International Series 1131*. Oxford: Archaeopress.

———., ed. 2006. *Ethnographies of Archaeological Practice: Cultural Encounters, Material Transformations*, Worlds of Archaeology. Walnut Creek: Left Coast Press.

European Association of Archaeologists. 2022. *Report: EAA 2022 Evaluation Survey*. Prague: EAA.

Everill, P. 2012. *The Invisible Diggers: A Study of British Commercial Archaeology*. 2nd Revised ebook edition. Available at https://www.academia.edu/4244219/The_Invisible_Diggers_A_study_of_British_commercial_archaeology_2nd_Edition_. Accessed 7 Feb 2023.

Favreau, J., and R. Patalano, eds. 2017. *Shallow Pasts, Endless Horizons: Sustainability & Archaeology: Proceedings of the 48th Annual Chacmool Conference*. Calgary: The Chacmool Archaeological Association of the University of Calgary.

Flatman, J. 2011. *Becoming an Archaeologist: A Guide to Professional Pathways*. Cambridge: Cambridge University Press.

———. 2015. 'A Slight Degree of Tension': Training the Archaeologists of the Future. *The Historic Environment: Policy & Practice* 6 (2): 142–155.

Fleck, L. 1979 [1935]. *Genesis and Development of a Scientific Fact*. Trans. F. Bradley. Chicago: University of Chicago Press.

Gould, R.A. 2000. Beyond Exploration: Underwater Archaeology after the Year 2000. *Historical Archaeology* 34 (4): 24–28.

Hamilakis, Y., and P. Duke, eds. 2007. *Archaeology and Capitalism: From Ethics to Politics*. London: Routledge.

Harmanşah, Ö. 2015. ISIS, Heritage, and the Spectacles of Destruction in the Global Media. *Near Eastern Archaeology* 78 (3) (Special Issue: The Cultural Heritage Crisis in the Middle East): 170–177.

Haslanger, S. 2018. What is Social Practice? *Royal Institute of Philosophy Supplement* 82: 231–247.

Heath-Stout, L. 2020. Who Writes About Archaeology? An Intersectional Study of Authorship in Archaeological Journals. *American Antiquity* 85 (3): 407–426.

Hodder, I. 1999. *The Archaeological Process: An Introduction*. Oxford: Blackwell.

———. 2010. Cultural Heritage Rights: From Ownership and Descent to Justice and Well-Being. *Anthropological Quarterly* 83 (4): 861–882.

Jackson, T. 2005. *Motivating Sustainable Consumption: A Review of Evidence on Consumer Behaviour and Behavioural Change*. Report to the Sustainable Development Research Network. Guildford: University of Surrey. Available at www.sustainablelifestyles.ac.uk/sites/default/files/motivating_sc_final.pdf. Accessed 7 July 2023.

Korea Cultural Properties Investigation and Research Institute Association. 2014. *Archaeological Heritage and the Public: Public Awareness and Conflict Management*. Seoul: Korea Cultural Properties Investigation and Research Institute Association and Cultural Heritage Administration of Korea.

Laluk, N.C., L.M. Montgomery, R. Tsosie, C. McCleave, R. Miron, S. Russo Carroll, J. Aguilar, Big wolf, A. Thompson, P. Nelson, J. Sunseri, I. Trujillo, G.M. DeAntoni, G. Castro, and T.D. Schneider. 2022. Archaeology and Social Justice in Native America. *American Antiquity* 87 (4): 659–682.

Latour, B. 1987. *Science in Action. How to Follow Scientists and Engineers Through Society*. Cambridge MA: Harvard University Press.

Little, B.J. 2007. *Historical Archaeology: Why the Past Matters*. Walnut Creek: Left Coast Press.

Lucas, G. 2001. *Critical Approaches to Fieldwork: Contemporary and Historical Archaeological Practice*. London: Routledge.

Marshall, Y. 2002. What Is Community Archaeology? *World Archaeology issue on "Community Archaeology"* 34 (2): 211–219.

McAtackney, L., M. Palus, and A. Piccini, eds. 2007. *Contemporary and Historical Archaeology in Theory: Papers from the 2003 and 2004 CHAT Conferences*, Studies in Contemporary and

Historical Archaeology 4. British Archaeological Reports International Series 1677. Oxford: Archaeopress.
McCann, D.P., and M.L. Brownsberger. 1990. Management as a Social Practice: Rethinking Business Ethics after MacIntyre. *The Annual of the Society of Christian Ethics* 10: 223–245.
McDavid, C. 2004. From 'Traditional' Archaeology to Public Archaeology to Community Action: The Levi Jordan Plantation Project. In *Public Archaeology as Applied Anthropology*, ed. P.A. Shackel and E.J. Chambers, 35–56. London: Routledge.
McGimsey, C.R. 1972. *Public Archaeology*. New York: Seminar Books.
———. 1984. The Value of Archaeology. In *Ethics and Values in Archaeology*, ed. E.L. Green, 171–174. New York: Free Press.
McGuire, R.H. 2008. *Archaeology and Political Action*. Berkeley CA: University of California press.
McManamon, F.P. 1991. The Many Publics for Archaeology. *American Antiquity* 65: 121–130.
Megill, K. 1968. Social Theory and Social Practice. *The Monist* 52(3), Philosophic Problems of Social Science, 452-464.
Merriman, N. 2004. Introduction: Diversity and Dissonance in Public Archaeology. In *Public Archaeology*, ed. N. Merriman, 1–18. London: Routledge.
Mickel, A. 2019. Essential Excavation Experts: Alienation and Agency in the History of Archaeological Labor. *Archaeologies* 15 (2): 181–205.
Moser, S. 2007. On Disciplinary Culture: Archaeology as Fieldwork and Its Gendered Associations. *Journal of Archaeological Method and Theory* 14: 235–263.
Moshenska, G., ed. 2017. *Key Concepts in Public Archaeology*. London: UCL Press.
Nilsson Stutz, L. 2018. A Future for Archaeology: In Defense of an Intellectually Engaged, Collaborative and Confident Archaeology. *Norwegian Archaeological Review* 51 (1–2): 48–56.
Olsen, B. 2010. *In Defense of Things: Archaeology and the Ontology of Objects*. Lanham MA: Altamira Press.
Orser, C.E., and Jnr. 1997. Professionalism in Historical Archaeology. *International Journal of Historical Archaeology* 1 (3): 243–255.
Perez-Alvaro, E. 2019. *Underwater Cultural Heritage: Ethical Concepts and Practical Challenges. Routledge Studies in Heritage*. London: Routledge.
Perkin, H. 1989. *The Rise of Professional Society: England since 1880*. London: Routledge.
Prescott, C. 2013. Archaeology vs. Science Or Taking Knowledge-Based Communication Seriously? *Current Swedish Archaeology* 21: 39–44.
Rankin, L. and Gaulton, G. 2021. Archaeology, Participatory Democracy and Social Justice in Newfoundland and Labrador, Canada. Archaeologies: journal of the World Archaeological Congress 17(1) (Archaeology as a Public Good), 79-102.
Reckwtiz, A. 2002. Toward a Theory of Social Practices: A Development in Culturalist Theorising. *European Journal of Social Theory* 5 (2): 243–263.
Renfrew, C. 2000. *Loot. Legitimacy and Ownership. Duckworth Debates in Archaeology*. London: Duckworth.
Roberts, J., and K. Sheppard. 2020. Introduction: Clusters of Knowledge. In *Communities and Knowledge Production in Archaeology*, Social Archaeology and Material Worlds, ed. J. Roberts, K. Sheppard, U.R. Hansson, and J.R. Trigg, 1–13. Manchester: Manchester University Press.
Roberts, J., K. Sheppard, U.R. Hansson, and J.R. Trigg. 2020. *Communities and Knowledge Production in Archaeology*, Social Archaeology and Material Worlds. Manchester: Manchester University Press.
Rubertone, P.E., ed. 2008. *Archaeologies of Placemaking: Monuments, Memories, and Engagement in Native North America*, One World Archaeology 59. Walnut Creek: West Coats Press.
Rudwick, M. 1985. *The Great Devonian Controversy: The Shaping of Scientific Knowledge Among Gentlemanly Specialists*. Chicago: University of Chicago Press.
Ruser, N., J. Leibold, K. Minro, and T. Hoja. 2020. *Cultural Erasure: Tracing the Destruction of Uyghur and Islamic Spaces in Xinjiang*, Policy Brief Report No. 38/2020. Canberra: Australian Strategic Policy Institute.

Rush, L., ed. 2010. *Archaeology, Cultural Property, and the Military*, Heritage Matters. Woodbridge: Boydell Press.

Sabloff, J.A. 2008. *Archaeology Matters: Action Archaeology in the Modern World*. Walnut Creek: Left Coast Press.

Saunders, N.J. 2004. Material Culture and Conflict: The Great War, 1914–2003. In *Matters of Conflict: Material Culture, Memory and the First World War*, ed. N.J. Saunders, 5–25. London: Routledge.

Scherzler, D., F. Siegmund, and P. Belford. 2019. Fokus: The Self-organization of Archaeologists in Europe. *Archäologische Informationen* 42: 147–148.

Schofield, J. 2010. Archaeology and Contemporary Society: Introduction. *World Archaeology* 42 (3): 325–327.

Schofield, J., J. Carman, and P. Belford. 2011. *Archaeological Practice in Great Britain: A Heritage Handbook*. World Archaeological Congress Heritage Manual Series. New York: Springer.

Shanks, M., and R.H. McGuire. 1996. The Craft of Archaeology. *American Antiquity* 61 (1): 75–88.

Sinclair, A. 2022 Archaeological Research 2014 to 2021: An Examination of Its Intellectual Base, Collaborative Networks and Conceptual Language Using Science Maps, *Internet Archaeology* 59. Available at: https://doi.org/10.11141/ia.59.10. Accessed 12 July 2023.

Skeates, R., J. Carman, and C. McDavid. 2012a. Introduction: Questioning Archaeology's Place in the World. In *The Oxford Handbook of Public Archaeology*, Handbooks in Archaeology, ed. R. Skeates, C. McDavid, and J. Carman, 1–10. Oxford: Oxford University Press.

Skeates, R., C. McDavid, and J. Carman, eds. 2012b. *The Oxford Handbook of Public Archaeology*, Handbooks in Archaeology. Oxford: Oxford University Press.

Smith, C. 2017. The Social and Political Sculpting of Archaeology (and Vice Versa). *Pyrenae: Revista de Prehistòria i Antiguitat de la Mediterrània Occidental* 48 (1): 7–44.

Smith, L., and E. Waterton. 2009. *Heritage, Communities and Archaeology. Duckworth Debates in Archaeology*. London: Duckworth.

Stone, P. 2006. 'All Smoke and Mirrors…' the World Archaeological Congress 1986–2004. In *A Future for Archaeology: The Past in the Present*, ed. R. Layton, S. Shennan, and P. Stone, 53–64. London: UCL Press.

Stone, P.G., ed. 2011a. *Cultural Heritage, Ethics and the Military*, Heritage Matters. Woodbridge, Boydell Press.

———. 2011b. Introduction: The Ethical Challenges for Cultural Heritage Experts Working with the Military. In *Cultural Heritage, Ethics and the Military*, Heritage Matters, ed. P.G. Stone, 1–28. Woodbridge: Boydell Press.

Stone, P.G., and J.F. Bajjaly. 2008. *The Destruction of Cultural Heritage in Iraq*, Heritage Matters. Woodbridge: Boydell Press.

Sullivan, W.M. 1995. *Work and Integrity: The Crisis and Promise of Professionalism in America*. New York: HarperBusiness.

Terrill, W.A. 2017. *Antiquities Destruction and Illicit Sales as Sources of ISIS Funding and Propaganda*. Carlisle Barracks PENN: Strategic Studies Institute, United States Army War College.

Thomas, S. 2012. Archaeologists and Metal-Detector Users in England and Wales: Past, Present, and Future. In *The Oxford Handbook of Public Archaeology*, ed. R. Skeates, C. McDavid, and J. Carman, 60–81. Oxford: Oxford University Press.

Thomas, S., C. McDavid, and A. Gutteridge. 2014. Editorial. *Journal of Community Archaeology and Heritage* 1 (1): 1–4.

Ucko, P.J. 1987. *Academic Freedom and Apartheid: The Story of the World Archaeological Congress*. London: Duckworth.

Van Reybrouck, D., and D. Jacobs. 2006. The Mutual Constitution of Natural and Social Identities During Archaeological Fieldwork. In *Ethnographies of Archaeological Practice: Cultural Encounters, Material Transformations*, Worlds of Archaeology, ed. M. Edgeworth, 33–44. Walnut Creek: Left Coast Press.

Van Velzen, D. 1996. The World of Tuscan Tomb Robbers: Living with the Local Community and the Ancestors. *International Journal of Cultural Property* 5 (1): 111–126.
Venclová, N. 2007. Communication Within Archaeology: Do We Understand Each Other? *European Journal of Archaeology* 10 (2–3): 207–222.
Walasek, H. 2020. Bosnia and the Destruction of Identity. In *Critical Perspectives on Cultural Memory and Heritage: Construction, Transformation and Destruction*, ed. V. Apaydin, 224–238. London: UCL Press.
Welch, D. 2017. Behaviour Change and Theories of Practice: Contributions, Limitations and Theories of Practice: Contributions, Limitations and Developments. *Social Business* 7 (3–4): 241–261.
Witmore, C., and M. Shanks. 2013. Archaeology: An Ecology of Practices. In *Archaeology in the Making: Conversations Through a Discipline*, ed. W.L. Rathje, M. Shanks, and C. Witmore, 380–398. London: Routledge.
Zimmerman, L. 2006. Liberating Archaeology, Liberation Archaeologies, and WAC. *Archaeologies* 2 (1): 85–95.

Chapter 8
Conclusion: Archaeology as Archaeology That Matters

The previous chapters have each discussed the various forms of practice that together make up the discipline and field of activity that is archaeology. The overall argument of this book—as stated at the outset in Chap. 1 and reiterated in other chapters—is that archaeology is important. A number of other commentators have expressed the same idea (see e.g. Witmore and Shanks 2013; González-Ruibal 2013; Smith 2017) and for all of us the reason for emphasising it, as stated in the closing sentences of Chap. 7, is the too frequent tendency of archaeologists to underplay what we do and the contributions we make to global society by a focus on only some aspects of what we do. While fieldwork (Chap. 2) is a distinctive feature of our field, ours is not the only fieldwork-based discipline. Other disciplines also produce outputs (Chap. 3) and engage in forms of communication (Chap. 4). Fewer others however have a role in serving the wider public through officially-supported and -sanctioned activity (Chap. 5), fewer still can produce direct political effects (Chap. 6) and while all disciplines are distinct communities, few are as fully social as archaeology needs to be in all it does (Chap. 7). But these are not why archaeology matters: it is the combination of these aspects that creates the significance of our field for the world. Beneath and around each of the themes covered in previous chapters are other themes—some expressly but partially addressed in other chapters, and others that are emergent and not so evident. These are the topics that will be addressed here to make the overarching argument of the book.

Archaeology as a Field of Ethics

As emphasised from the beginning, a running theme of the book is that of ethics. Highlighted specifically here and there throughout, it has also lurked below the specifics of individual topics. Ethics, as indicated previously, are not easy to engage with: more than one ethical option may be available and none may be more worthy

J. Carman, *Archaeological Practice as Politics and Ethics*, https://doi.org/10.1007/978-3-031-69828-6_8

than the other in abstract terms. Individual circumstances must each be approached in their own terms, and while there may be general principles acting as guidance, what should be *done* or *not done,* or *said* or *not said,* in a particular context will depend upon the specifics of that case. This means that all archaeological decisions are made under the aegis of ethics: none are merely the application of practical responses to practical problems, but at the same time none are straightforward and all are necessarily pragmatic.

Ethical issues underlie any attempt at a definition of what is archaeology and who can be considered an archaeologist (Chap. 1): formally, in terms of the degree of professionalisation and adherence to closely defined disciplinary standards required; and less so in terms of how the work and those conducting it affect the world around them. It is in some quarters considered more ethically sound to demand some form of formal qualification of all archaeologists, to ensure that they know and appreciate the standards to which they need to work and comply; but this would exclude those who have had fewer opportunities for formal education but who may be perfectly effective and just as likely (or possibly more so) to meet high moral standards—such as members of Indigenous or other subaltern groups who can inject an understanding of the effect on them of the appropriation of their pasts by others. The appropriation of others' pasts by select groups—such as archaeologists employed by the state or private agencies—may itself be seen as unethical (e.g. Scarre and Coningham 2013), while at the same time allowing exclusionary control over that past by a stakeholder community is a denial of the more general ethical principle that "the past belongs to all" (Merriman 2004, 1).

The arena of fieldwork (Chap. 2) is fraught with ethical concerns. Archaeology is all too frequently the retrieval of material from the earth—where it may have lain for thousands of years, much of which was deposited deliberately with no intention of retrieval (as in cemetery sites, for example). If archaeologists accept the role of stewards of that material, it may be more ethically appropriate to leave it where it lies than remove it for investigation; on the other hand, to abandon the practices of investigation of this material is to abrogate archaeology altogether and may be to abandon culturally significant material to loss and destruction. As an arena of training, responsibilities in fieldwork apply to the material remains of the past—which is subject to discovery and retrieval by those less- or untrained in the techniques of such work—and to those subject to training who are our students and future colleagues. We are bound to treat these others with respect and ensure their welfare, especially in addressing issues of abuse by those in positions of authority over them—an issue of increasing significance (Coltofean-Arizancu et al. 2023). The decision has always to be made whether excavation (and thereby the destruction—or as Lucas [2001, 40] has it, 'alteration'—of the site) is required or justified, or whether some kind of non-invasive survey and recording technique is more appropriate, especially in dealing with sensitive pasts and communities (see e.g. Sturdy Colls 2015). Chapter 2 's final section on Exploration emphasised the link between the exploration by Europeans of others' territories and their conquest and colonisation by European states: the implication of archaeologists in the European imperial project is itself an issue of ethics and moral choice and subject to increasing study and critique (Diaz-Andreu 2007).

Producers (Chap. 3) owe duties to the material with which they work and the audiences for the products of that work. In relation to material, archaeologists owe a duty of care to ensure that remains of the past are not damaged or destroyed to no purpose—a duty deriving from the status as steward—and that what we say about that material and those who produced it in the past are dealt with fairly in our interpretations. The act of creation of outputs raises issues of ethics in terms of how we engage with those who have a claim on the past under investigation—whether as a local or descendant community (Marshall 2002, 216)—and especially those who suffer under some form of social disadvantage. It also raises questions relating to any form of cultural production—how much credit should be given to colleagues and predecessors for the outcomes of any particular piece of work. Here we encounter one of the underlying questions at the core of this book—who (or what) is archaeology *for* and whose interests does it serve? It causes archaeologists to engage with serious issues relating to the world in which our work is conducted and inevitably engages with issues that are essentially political in nature (and see below). These are moral choices not practical ones: the choices are driven not by the methods and techniques of investigating the past, but by more fundamental ideas about who has authority to determine what the past was like and how it resonates—or not—with presents and futures. It determines how individual archaeologists will work with or for communities, and what kinds of communities. The ethical issue here is what kind of person an archaeologist should be and where the duties lie.

As communicators (Chap. 4) all of the above concerns are reflected in how and what we communicate and to whom. Our relations with colleagues, rivals and successors are all implicated here (see e.g. Connah 2010). While many (If not all) archaeologists will accept that scientific knowledge, including archaeological knowledge, belongs to the world, and while work is more and more frequently offered in Open Access form as increasingly required by funding and employing institutions, it is always a question as to how much of one's work will be made available to others through publication and at conferences. Apart from the fear of challenge to one's findings (a nervousness with which we all live to a greater or lesser extent), there is the fear of plagiarism and the misuse of one's work for others' purposes (see e.g. the use by Holocaust deniers of archaeological work at Treblinka: Sturdy Colls 2015, 348–349). In similar vein, there may be a need to retain some information to prevent damage to a site that may be attractive to mass visitors, looters and collectors (McCarthy and Wickert 2022). It is interesting to note how nervous we tend to be when discussing in closed professional contexts among ourselves any aspect of our own practice that goes beyond positive reinforcement: especially discussions of our engagements with those outside the discipline. In my experience, these are always described in positive terms, rarely involve any poor practice, and no community project has ever achieved anything except good results, even though any results do not survive the project itself. There is an ethical issue here too: that of being honest with each other about our work and its consequences.

As a public service (Chap. 5) archaeology also engages with ethical requirements that go beyond mere practice. As a public service, the field is obviously fulfilling a need for a wider community and it is the needs of that community which determine how the work is conducted. There are the ethical requirements common to any such provision—to do the work effectively and to cost and time and in recognition that the work ultimately belongs not to the archaeologist but to the contractor or employer. At the same time, the ethical issues that drive fieldwork and other aspects of the field are also operative: duties are owed to the material, its prospective audiences, the community whose past is under investigation, and to colleagues, employees and any volunteers also involved. Where archaeology is conducted as part of government provision the requirements become more pressing, raising issues relating to the political uses to which the work may be put: archaeological work has been part of projects to displace populations, to assert control by one group over another, to allow occupation by invaders and colonisers, and as cover for intelligence gathering and other explicitly military purposes (Hamilakis and Duke 2007). Archaeologists conducting work on behalf of official agencies—governmental or otherwise—carry authority that goes beyond a concern for interpreting pasts: they control and can demand the allocation of resources that may be used for other purposes; they may indeed represent the full power of an authority which cannot be challenged without serious consequences for any prospective challenger of that authority. All of this places a heavy moral burden on us.

Archaeology is—as emphasised in several chapters and especially Chap. 6—inevitably a political field. It operates at the service of political entities such as governments. It acts frequently as an agent of governance, commanding resources from the wider community. It is used to support structures of authority—especially the nation state as the most common form of political organisation — and provides justifications for them in the form of deep time-depth. It is used similarly to uphold social norms such as gender characterisations and functions, or ideas of family and household. But it also serves to challenge these ideas by providing alternative ideas of how the world has been and could be. It can reveal the unrecognised manners of control and dominance at work in our own modern societies and explain how they came into being. Archaeologists draw upon this to engage directly in political activity, working with disadvantaged communities and the oppressed to challenge authority. They work in areas of conflict to ameliorate the consequences of violence, and to help rebuild communities broken by division and war by restoring evidences of common pasts. Archaeology is used by states to raise their status in the world and make friends and allies of other states (Luke and Kersel 2013). All these applications of our activities, our material and our interpretations carry with them important moral and ethical questions, especially when we allow others to use archaeology for their purposes without any control by us.

It is evident from every chapter and the foregoing that archaeology is inevitably a social practice (Chap. 7). It is this in two senses: archaeology as a discipline is a product of the kind of societies we have created and inhabit; but it is also conducted in groups, and there is arguably no such thing as a lone archaeologist because all our work is conducted as part of the wider community of archaeologists. Many archaeologists consider themselves to be professionals, and there is more to professionalism than formal qualification and payment for services: it carries with it adherence to rules and norms of behaviour and quality of performance. At the same time, archaeologists—professional in this narrow sense and otherwise—come together at conferences and form various kinds of associations. Collectively we determine who can be considered 'one of us' and those who cannot: we divide others into enemies, allies and neutrals (Table 7.1; Carman 2015, 178–180). To what extent we involve ourselves in these structures of disciplinary organisation is very much a moral and ethical choice: to choose non-involvement is in itself to make an ethical decision because these are all the defining characteristics of our discipline. To be an archaeologist is to choose to be a professional or not, to engage with others or not, to define others as a fellow archaeologist or not, and the kinds of relationships to be had with those others.

It is clear that there is no escape from issues of ethics in archaeology, a point that may be surprising to those who consider that archaeology is only about understanding lost pasts. All our work has implications in the present, where it is conducted and where its effects are felt. This is one reason why archaeology matters: everything every archaeologist does—recognised or otherwise—has resonances for today and tomorrow. To be an archaeologist is to engage with important ethical issues at every turn—many of which we do not see because they are built in to the very structure of the discipline.

Archaeology as a Decolonising Endeavour

Archaeology is one among many fields currently concerned to 'decolonise' itself. Decolonisation is the process whereby a field of activity first identifies its role in the colonising of others and proceeds to divest itself of Western dominance, allowing alternative voices and alternative modes of practice to emerge. As such, it is inevitably closely related to the field of ethical choices, since the decisions as to what alternative voices and practices can be considered legitimate and those that are not are in essence ethical decisions (as discussed above); they are also closely related to archaeology as a realm of politics (Chap. 6). The term appears once only in the text of this book (in Chap. 6)—referring especially to Smith's understanding of the process as one "characterised by a concern with rethinking colonial histories…, deconstructing power relations between Indigenous peoples and archaeologists…, shared histories… and shaping archaeological practice so that it benefits Indigenous communities" (Smith 2017, 9–10), but the idea necessarily runs through discussions in all previous chapters.

There is of course more to decolonisation than reshaping relations with Indigenous peoples, because there are significant other communities that have been subject to colonisation, as Langer and Matić (2023) indicate for archaeologies of Egypt and Sudan. They distinguish 'decolonisation'—which is the political process of handing government to local élites—from 'decoloniality' which "aims to overcome coloniality and create a world free of any colonial power [political, intellectual or otherwise], realized in the form of the *pluriverse*" (Langer and Matić 2023, 15; emphasis in original). A pluriverse recognises that only are there different worldviews but, beyond that, entirely different worlds inhabited by diverse communities. As such, a process of decolonisation in these terms engages not only with the effects of past and present imperialism, but also systems of identity construction and gender relations imposed in Western terms. This therefore also connects to the development of intersectionality studies (mentioned in Chaps. 3 and 6) whereby different aspects of disadvantage—gender, ethnicity, religion, social class, disability and geography—are treated together rather than as entirely separate strands of experience (e.g. Spencer-Wood and Trunzo 2022).

Examples of efforts to achieve the decolonisation of archaeology as a discipline occur in every chapter. A concern for the communities—especially those considered as subaltern, whether Indigenous or otherwise—is cited as included in Codes of Practice, as noted in Chap. 1 and also emerges in aspects of fieldwork across the globe, as noted in Chaps. 2 and 3. The role of archaeologists in the colonisation of others is highlighted especially in the section on exploration in Chap. 2. The discussion in Chap. 4 of archaeology as a system of communication (Mizoguchi 2006) indicated efforts to include the voices of those otherwise excluded from archaeological discourse in education (e.g. Quave et al. 2021) and to include in such education specific reference to the complexity of kinship relations of Indigenous groups to challenge and overcome assumptions of 'simplicity' in such societies derived from Western thinking (Smith and Burke 2007). The same chapter also introduced ideas about the sharing of ownership of archaeological work and its products which serves to challenge Western-imposed models of the authority given to expertise, also highlighted in Chap. 5 in the discussion of how systems of archaeological heritage management serve to deny control over cultural resources to the communities who produced them.

Chapter 6 examined specifically how subaltern, and especially Indigenous, groups are controlled and governed by the exercise of expertise in a Western model of scholarly activity, described by Hutchings and La Salle (2017) as a "State Heritage Crime". Others examine how archaeology serves to deny economic power to communities (e.g. Lafrenz Samuels 2009) and others again how the study of archaeologies of the contemporary past and present helps understanding and developing responses to various forms of social deprivation and exclusion (Smith 2017). The idea proposed by McGuire (2008, 16) that "archaeology is politics by other means" is also an aspect of decolonisation, since Western models of scholarly expertise specifically deny any political role: to recognise our inevitable involvement in the political is the first step towards engagement with the legacies of colonialisation. In such a mode, Chap. 7 mentions the formation of WAC as a body dedicated to "liberate archaeology from its practice of scientific colonialism" (Zimmerman 2006, 87) and Nilsson Stutz's (2018), 54–55) argument that archaeology is "ideally situated to lead debates about cultural heritage, identity politics, postcolonialism, migration", etc.

The task of decolonising archaeology is by no means complete, is frequently a case of one step forward and two back, and not all branches of the field are yet fully engaged. Nonetheless, whatever we call it ('decolonisation', 'archaeology as politics', 'democratising archaeology' are some alternatives) it is a growing factor in the archaeological endeavour and one of which we can—like Nilsson Stutz—be proud to be associated.

Archaeology as a Field of Temporalities and Physicalities

Chapter 3 discussed the engagements of archaeologists with philosophical issues, and especially those related to issues of temporality and materiality. Archaeology, although notionally the study of pasts, is in fact an encounter in the present with material which is then analysed and interpreted for an audience in the future: as suggested in Chap. 3 archaeology is about the production not only of pasts but also of futures and this is done in the present. Moreover, there is a growing trend in archaeology for the study of our own present (e.g. Harrison and Schofield 2010) which firmly locates archaeology among those fields that study our own social context. As Greenberg (2023, 1787) puts it, this "expansion of archaeology to new places, times and people is an opportunity to enlarge our discursive field and find new intellectual partners, within academia and outside it; partners who are intrigued not so much by antiquity, but by how archaeologists see".

Such developments inevitably raise questions concerning what kind of discipline archaeology is or should be. As discussed in Chap. 1, some argue that "Archaeology is anthropology or it is nothing" (Willey and Phillips 1958, 2), others that archaeology is a form of long-term history-making (Hodder 2010), and yet others simply that "archaeology is archaeology is archaeology" (Clarke 1968, 12–13) and a discipline in its own right. It is really a matter of the tradition of archaeology in which we work that determines our response to these different claims. As Cruz Berrocal (2012) points out, the connection with anthropology is most strongly made in North American archaeology, where archaeology is one of the four components of anthropological study (along with physical anthropology, cultural anthropology and linguistics). From a Spanish perspective, and focusing upon the study of prehistory as a dominant strand and especially the Iberian tradition of drawing inspiration from Marxism, Cruz Berrocal argues that:

> We do not have to choose between being anthropologists or historians…. But from the point of view of the critical study of social change and resistance, dealing with contingency, attention to the difference, and taking critical looks at our notion of progress, archaeology is history or it is nothing (Cruz Berrocal 2012, 43).

This understanding of archaeology unites the concern with the past with critiques of the present—reflecting the political nature of archaeology and especially with progressive and radical perspectives and aims, which in turn feed into the construction of envisaged futures.

Apart from its ethical and political aspects, archaeology is also—some might argue predominately—an encounter with the physical world. It is this in two senses: we deal with the material residues of past human activities (Lucas 2001; Olsen 2010) to understand the world around us; and in doing so archaeology as fieldwork is an intensely *physical* activity involving the skilful use of tools (Edgeworth 2006, 38). One of the distinctive qualities of archaeology as an academic discipline is indeed this combination of the practical with the cognitive: it is one of the things that makes archaeology attractive to students and volunteers. It also provides the basis for multiple routes of entry to the field: those less suited to formal academic study may nonetheless find their way into the field through a route based on the acquisition of more practical skills (e.g. apprenticeships: Nash and James 2023). This inevitably places the focus on fieldwork as a distinctive component of archaeology as a discipline, and on technical accomplishment rather than abstract cognitive ability. Nonetheless, as a number of commentators note (e.g. Hodder 1999), the practice of fieldwork is also a process of interpretation, involving the unity of thought and physical experience: the changes in soil and its content that signal the presence of material of interest are as much felt through the fingers as seen; and the physical sensation precedes the interpretive response.

That engagement with the material occurs in the present: while the 'stuff' may represent a past which we try to understand, what we do to interpret that material happens now. This manages to combine the multiple temporalities in which archaeology is entangled with the essentially material nature of the work. This is reflected not only in the nature of fieldwork but also in the physical products of archaeological interpretation—written such as books and articles, constructed such as exhibitions, preserved in the form of upstanding sites and monuments, and other tangible outputs in terms of those from work in the public sector as a contribution to planning and development, all of which are part of the *now*. Archaeologists' encounters and engagement with each other and others—in the form of meetings and outreach activities—are similarly current activity: although supposedly *about* the past they are not *past* in themselves (as reflected in our own understandings of how the field has developed: Rathje et al. 2013). As discussed in Chap. 3, heritage work too is very much about futures rather than pasts or even presents (Harrison 2016) while Borck (2018) argues that archaeology as heritage work (and as a form of 'chronopolitics') is used to legitimise present social forms and construct futures that deny alternatives. Here all the practices of archaeology—those involved in fieldwork, in the production of outputs and outcomes, in our modes of communication, as a public service, as an engagement with politics, and in our relations with each other and others—are united in a direct engagement with material and temporality.

All this feeds into the argument of this book that archaeology is important. It is important because it unites the material and the intellectual and these with ideas about pasts, presents and futures and the alternatives they represent in terms of modes of existence. The past offers alternative ways of being that cannot be dismissed as impractical because they have been applied in practice; the past also serves to explain how and why our world is as it is now and to demonstrate that things could be different; and questioning presents opens the door to the creation of futures that are not mere replications of the present. While other related disciplines—history, anthropology, sociology etc.—also do some of these, it can be argued that none do all of these, as archaeology can and does. Too often, what we do not do is justify ourselves strongly enough.

Archaeology as a Global Endeavour

It is interesting how little we emphasise the truly global—and indeed more than global—nature of our discipline. There is archaeology everywhere where humans have lived or visited—including such inhospitable places as the arctic and Antarctica (Senatore 2019), under the oceans (Perez-Alvaro 2019), and on the moon and elsewhere in outer space (Gorman and O'Leary 2013)—and archaeology that indicates the entire history of our species. There are archaeologists working in every country of the globe: some are quite well-endowed with archaeologists, especially in the global North, while others can boast only a few, especially poorer countries in the global South. Every country in the world—including those classed as 'failed' states—has a body of law governing its archaeology, the uses to which it can be put, and in whose control it lies (O'Keefe and Prott 1984). Our conversations with each other are frequently conducted across period and area interests and—as indicated by the existence of organisations such as IUPPS and WAC—can also be global (and see the inclusion of commentators from other regions in texts dedicated to particular regions: e.g. Kim and Zoh 2023). In short, archaeology and archaeologists are a universal phenomenon. What is more, and this provides a contrast with other disciplines, an archaeologist from one part of the world can meaningfully discuss disciplinary issues with any other archaeologist regardless of geographical base or period or regional interest: one reason is shared methodology but also underlying purpose and the knowledge of archaeology's global reach.

The concerns of archaeologists are equally global. Huvila et al. (2022) list a number of contemporary issues of global concern on which archaeologists may provide insights, among them "social inequality, gender, norms, acceptable choices, migration, environmental exploitation, ecological and social sustainability" as well as conflict (Huvila et al. 2022, 298–300). Similarly, Ribeiro and Giamakis (2023)—echoing concerns of McGuire (2008) and others — emphasise the persistent effects of class distinctions as a direct concern for the field. What is notable in such discussions, however, is that these authors are less concerned with the world external to archaeology than with the discipline of archaeology itself. Huvila et al. (2022) point to "the constant importance of continuing and developing the interdisciplinary effort of investigating archaeological practices and knowledge work as a reflexive intra disciplinary effort within archaeology itself and by conducting studies sideways, up, and down from different disciplinary and theoretical perspectives to develop a better understanding of the limits and opportunities of archaeological knowledge in the contemporary and future society" (Huvila et al. 2022, 303). Ribeiro and Giamakis (2023) are similarly less concerned with how archaeologists study class relations than—as their title makes clear—"class and elitism *in archaeology*" (emphasis added). As they put it in their conclusion, archaeologists need to be:

> more critically aware of personal privileges, whether it was financial help from parents or whether an academic appointment came via a favour …; it is also important to recognise the work being done in marginalised contexts…; also, one should pay attention to biases when conducting reviews or hiring new staff…. These, and countless other little habits, are the source of a better world. Will these small changes counteract centuries of classist and elitist build-up that sustains archaeology? Obviously not, but what they will do is create a discipline that is considerably less arrogant and more inclusive, one that we can certainly be less ashamed of (Ribeiro and Giamakis 2023, 18).

It will be clear that the position of this book is that involvement in archaeology is not a matter for shame: but the points made by these authors and others about the need for introspection by archaeologists are both well made and representative of what so many in the field are already engaged in. As noted in Chaps. 1 and 7 and implied above, archaeological organisations across the globe have taken pains to set out the minimum standards of behaviour for their members (see Table 1.1) and these have then been subject to critique by those who wish to set a higher ethical tone that goes beyond narrow professionalism and responsibility to paymasters (see e.g. Zimmerman 1995; Tarlow 2001; McGuire 2012). As noted in Chap. 3 and elsewhere, there has been a particular concern to accommodate the interests of Indigenous and other descendant communities as a result of growing recognition that their pasts as represented by material remains are not the property of archaeologists but the community whose past the material represents. Beyond that, there are drives to recruit to the discipline from Indigenous and descendant communities as well as other less advantaged groups to give a voice in the field to those of different experience and thereby a more diverse community of archaeologists. Although not entirely unique, and as mentioned in Chap. 7, the work of WAC in this area is especially notable (Ucko 1987; Day 2006; Stone 2006) and has been influential in encouraging other organisations to recognise the status of Indigenous peoples in relation to archaeology. Such concerns for the local in relation to archaeology have become a global force over the past few decades and will continue into this twenty-first century.

It matters that archaeology is understood to be a global endeavour rather than as a set of parallel local or national efforts. If we accept the unity of humanity—an idea that emerged centuries ago—then the study of all of the human past as offered by archaeology, going beyond as it does the existence of textual evidence, is also a unity. The big themes of human existence over the long term and changes in the way human societies organise themselves and others, the effects of environment, and of human activity on environment, of how human communities interact with each other for good or ill, and how we became what we are and how we engage with the world we have created for ourselves—all these and more are issues of global relevance. That they play out differently in different places in different times is not a concern: that is no argument for disunity among archaeologists, because ultimately our interests centre on the same themes. That we may encounter difference and disagreement among ourselves is a strength and it is in engaging with those differences that the unity of archaeology lies. To rephrase others cited above (Willey and Phillips 1958, 2; Cruz Berrocal 2012, 43) and turn their rhetoric to my purpose: archaeology is *global* or it is nothing.

Archaeology as a Discipline in Flux

A brief, highly reductive and in many ways inaccurate timeline for archaeology as a discipline may read as follows:

Fifth century BCE to eighteenth century CE	Antiquarian interests in recording monumental pasts
Eighteenth century to twentieth century CE	Development of fieldwork methodologies
	Increased focus on buried remains
	Archaeology as a scientific study of prehistory
	Archaeology as a tool of European imperialism
Mid-twentieth century CE	Scientific techniques allow firm dating
Mid- to late twentieth century CE	Debates about disciplinary links
Late twentieth century CE	Debates about purpose of archaeology
Twenty-first century CE	Archaeology as a field of political engagement

It is worth noting that archaeology has always been conducted across territorial boundaries, and that archaeologists and their antiquarian predecessors have communicated across those boundaries so the essence of the global nature of the field was present from its inception. It became truly global as a result of the colonisation of others' territories by Europeans and particularly as the nation state became established as the universally-applicable form of political association. As noted in Chap. 6, the connection between the nation state and archaeology is no accident: the new form of political organisation required a deep past to justify itself and antiquarian study of material pasts could provide this (see e.g. Diaz-Andreu 2007). Archaeologists have since become more critical of their own field and its role in history: although in many ways still related to state activity (see Chap. 5) and as noted above, archaeologists also seek to work with and for the disadvantaged whose pasts are under study. This does not mean there are not those who operate in a more antiquarian manner than scientific, or who do not consider any period except the prehistoric as 'archaeological', or do not recognise archaeology as inevitably a field of politics, so older understandings of the field remain in place for some. Nevertheless, archaeology has never remained fixed at a certain point in its development and continues to change as the world changes.

The role of intra- and inter-disciplinary debate—as issue raised especially in Chap. 4—is crucial here. It is because archaeology never stands still, not only in terms of theoretical and methodological development but also in terms of its contested purposes, that debates between archaeologists and engagements with others with interests in the pasts under study are at once required and conducted. As

Kristiansen (2009, 3) has noted "During the last generation archaeology has expanded its boundaries as a discipline in many new directions, which makes it difficult to define as a homogenous discipline". He goes on to suggest that "Situated in the dialectic between past and present, knowledge and interest, science and the humanities, preservation and excavation, research and public presentation, archaeology remains one of the most challenging and wide-ranging academic disciplines to engage in" (Kristiansen 2009, 38). For Harris and Cipolla (2017) the way forward for archaeological theory in the twenty-first century is the denial of dualistic thinking and a concern with the complexities of 'entanglement' which also have implications and consequences for archaeological practice. Here is where the physical activities of archaeologists—including in the field—and their intellectual processes (never very far apart as I hope previous chapters have demonstrated) come together: theory not only influences practice, but practice also affects theory, which is where the philosophising of archaeologists (Chap. 3) plays its part. What we do encourages thought and interaction with others, and thought and interaction with others encourages new activity. The cycle goes on and archaeology as a discipline responds by changing.

This ability to constantly adapt and to incorporate not only new method and theory but also new topics, ranging in period from the earliest emergence of humanity to the present, together with the future-orientation of heritage work and political activism, is what marks archaeology as significant and crucially as different from other fields. Together with its global coverage and inevitably underlying ethical nature, the capacity for change and development in the field make archaeology a noble enterprise, a point we frequently are too slow to acknowledge. The argument of this book is that archaeology is important: and while others cited throughout his book point to specific issues on which archaeology has something important to say (see e, g. Little 2007; Sabloff 2008), they too often miss the larger canvas on which archaeology operates. As this book has sought to emphasise, archaeology does not only tell us specific things about particular pasts: it works in the present to inform and indeed guide the construction of futures. Moreover, archaeology—despite its popular branding, including too often by ourselves—is not exclusively about retrieving and interpreting things from the past: it is a locus for the production of ideas (Chap. 3) and an arena for various modes of communication (Chap. 4), a public service (Chap. 5), a realm of political debate and action (Chap. 6), and overarching all this a highly social environment in which to operate, in every sense (Chap. 7).

As archaeologists, we have contributions to make on discussions and decisions concerning all the big issues that affect the world, and we do so: and we can be and are an arena for promoting change for the betterment of all of humanity through our global reach and willingness to take ethical positions. It cannot be said too often that our discipline covers all human pasts and everywhere impacted by human activity: we are not limited by place nor much by time. That makes what we do matter and it makes us matter too. The world needs archaeology and archaeologists, and we are required to recognise this for ourselves.

References

Borck, L. 2018. Constructing the Future History: Prefiguration as Historical Epistemology and the Chronopolitics of Archaeology. *Journal of Contemporary Archaeology* 5 (2): 213–302.

Carman, J. 2015. *Archaeological Resource Management: An International Perspective*. Cambridge: Cambridge University Press.

Clarke, D.L. 1968. *Analytical Archaeology*. London: Methuen.

Coltofean-Arizancu, L., B. Gaydarska, S. Plutniak, L. Mary, M. Hlad, I. Algrain, B. Pasquini, S. Vandevelde, E. Stamataki, P. Janežič, B. Wouters, and A. Sengeløv. 2023. Harassment, Assault, Bullying and Intimidation (HABI) in Archaeology: A Europe-Wide Survey. *Antiquity* 97 (393): 726–744.

Connah, G. 2010. *Writing About Archaeology*. Cambridge: Cambridge University Press.

Cruz Berrocal, M. 2012. Archaeology Is (Sometimes) History or it is Nothing: The Value of History as Critical Archaeology. In *The Prehistory of Iberia: Debating Early Stratification and the State*, ed. M. Cruz Berrocal and A. Gilman, 29–49. London: Routledge.

Day, M. 2006. Peter Ucko and the World Archaeological Congress (WAC). In *A Future for Archaeology: The past in the Present*, ed. R. Layton, S. Shennan, and P. Stone, 41–46. London: UCL Press.

Diaz-Andreu, M. 2007. *A World History of Nineteenth Century Archaeology: Nationalism, Colonialism and the Past*. Oxford: OUP.

Edgeworth, M. 2006. *Acts of Discovery: An Ethnography of Archaeological Practice*, British Archaeological Reports International Series 1131. Oxford: Archaeopress.

González-Ruibal, A., ed. 2013. *Reclaiming Archaeology: Beyond the Tropes of Modernity*. London: Routledge.

Gorman, A.C., and B.L. O'Leary. 2013. The Archaeology of Space Exploration. In *The Oxford Handbook of the Archaeology of the Contemporary World*, ed. P. Graves-Brown, R. Harrison, and A. Piccini, 409–424. Oxford: Oxford University Press.

Greenberg, R. 2023. More Than Antiquity: How Archaeologists See. In Ben-Yosef, E., Jones, I.W.N. (eds) *"And in length of Days Understanding" (job 12:12). Interdisciplinary Contributions to Archaeology*. New York: Springer. https://doi.org/10.1007/978-3-031-27330-8_75. Accessed 27 Sept 2023.

Hamilakis, Y., and P. Duke, eds. 2007. *Archaeology and Capitalism: From Ethics to politics*. London: Routledge.

Harris, O.J.T., and C. Cipolla. 2017. *Archaeological Theory in the New Millennium: Introducing Current Perspectives*. London: Routledge.

Harrison, R. 2016. Archaeologies of Emergent Presents and Futures. *Historical Archaeology* 50 (3): 165–180.

Harrison, R., and J. Schofield. 2010. *After Modernity: Archaeological Approaches to the Contemporary Past*. Oxford: Oxford University Press.

Hodder, I. 1999. *The Archaeological Process: An Introduction*. Oxford: Blackwell.

———. 2010. Cultural Heritage Rights: From Ownership and Descent to Justice and Well-being. *Anthropological Quarterly* 83 (4): 861–882.

Hutchings, R.M., and M. La Salle. 2017. Archaeology as State Heritage Crime. *Archaeologies* 13 (1): 66–87.

Huvila, I., C. Dallas, M. Toumpouri and D. Ní Chíobháin Enqvist. 2022. Archaeological Practices and Societal Challenges. Open Archaeology 8, 296–305. https://doi.org/10.1515/opar-2022-0242. Accessed 3 Oct 2023.

Kim, J., and M. Zoh, eds. 2023. *Asia's Heritage Trend: Examining Asia's Present Through Its Past*, Routledge Contemporary Asia Series. London: Routledge.

Kristiansen, K. 2009. The Discipline of Archaeology. In *The Oxford Handbook of Archaeology*, ed. C. Gosden, B. Cunliffe, and R.A. Joyce, 3–46. Oxford: Oxford University Press.

Lafrenz Samuels, K. 2009. Trajectories of Development: International Heritage Management of Archaeology in the Middle East and North Africa. *Archaeologies* 5 (1): 68–91.

Langer, C., and U. Matić. 2023. Postcolonial Theory in Egyptology: Key Concepts and Agendas. *Archaeologies* 19 (1) Special Issue on Postcolonial Theory in Egyptology: Applications, Debates and Potentials, 1-27: 1–27.

Little, B.J. 2007. *Historical Archaeology: Why the Past Matters*. Walnut Creek: Left Coast Press.

Lucas, G. 2001. *Critical Approaches to Fieldwork: Contemporary and Historical Archaeological Practice*. London: Routledge.

Luke, C., and M. Kersel. 2013. *U.S. Cultural Diplomacy and Archaeology: Soft Power, Hard Heritage*. London: Routledge.

Marshall, Y. 2002. What Is Community Archaeology? *World Archaeology issue on "Community Archaeology"* 34 (2): 211–219.

McCarthy, J.P., and J. Wickert. 2022. On the Horns of an Archaeological Dilemma: Balancing Site Confidentiality and Public Interpretation Imperatives at Delaware State Parks. In *Creating Participatory Dialogue in Archaeological and Cultural Heritage Interpretation: Multinational Perspectives*, ed. J.H. Jameson and S. Baugher, 67–83. New York: Springer.

McGuire, R.H. 2008. *Archaeology and Political Action*. Berkeley: University of California Press.

McGuire, R. H. 2012. Critical Archaeology and Praxis (Kritische Archäologie und Praxis). Forum Kritische Archäologie 1: 77–89. https://doi.org/10.6105/journal.fka.2012.1.11. Available at http://www.kritischearchaeologie.de/repositorium/fka/2012_1_11_McGuire.pdf. Accessed 3 Oct 2023.

Merriman, N. 2004. Introduction: Diversity and Dissonance in Public Archaeology. In *Public Archaeology*, ed. N. Merriman, 1–18. London: Routledge.

Mizoguchi, K. 2006. *Archaeology, Society and Identity in Modern Japan*, Cambridge Studies in Archaeology. Cambridge: Cambridge University Press.

Nash, A., and A. James. 2023. *Apprenticeships in the Historic Environment Sector: Examining Employer Interest and Barriers to Implementation*, Research Report Series 43/2023. Portsmouth: Historic England.

Nilsson Stutz, L. 2018. A Future for Archaeology: In Defense of an Intellectually Engaged, Collaborative and Confident Archaeology. *Norwegian Archaeological Review* 51 (1–2): 48–56.

O'Keefe, P.J., and L.V. Prott. 1984. *Law and the Cultural Heritage: Volume 1: Discovery and Excavation*. Abingdon: Professional Books.

Olsen, B. 2010. *In Defense of Things: Archaeology and the Ontology of Objects*. Lanham: AltaMira Press. Available at ProQuest Ebook Central, https://ebookcentral.proquest.com/lib/bham/detail.action?docID=616256. Accessed 17 Mar 2023.

Perez-Alvaro, E. 2019. *Underwater Cultural Heritage: Ethical Concepts and Practical Challenges*, Routledge Studies in Heritage. London: Routledge.

Quave, K.E., S.M. Fie, A.-S.Q.Q. Greiff, and D.A. Agnew. 2021. Centering the Margins: Knowledge Production in the Introductory Archaeology Course. *Advances in Archaeological Practice* 9 (2): 87–100.

Rathje, W.L., M. Shanks, and C. Whitmore, eds. 2013. *Archaeology in the Making: Conversations Through a Discipline*. London: Routledge.

Ribeiro, A., and C. Giamakis. 2023. On Class and Elitism in Archaeology. Open Archaeology 9, 1–22. https://doi.org/10.1515/opar-2022-0309. Accessed 3 Oct 2023.

Sabloff, J.A. 2008. *Archaeology Matters: Action Archaeology in the Modern World*. Walnut Creek: Left Coast Press.

Scarre, G., and R. Coningham, eds. 2013. *Appropriating the Past: Philosophical Perspectives on the Practice of Archaeology*. Cambridge: Cambridge University Press.

Senatore, M.X. 2019. Archaeologies in Antarctica from Nostalgia to Capitalism: A Review. *International Journal of Historical Archaeology* 23: 755–771.

Smith, C. 2017. The Social and Political Sculpting of Archaeology (and Vice Versa). *Pyrenae: Revista de Prehistòria i Antiguitat de la Mediterrània Occidental* 48 (1): 7–44.

Smith, C., and H. Burke. 2007. The Skin Game: Teaching to Redress Stereotypes of Indigenous People. In *Archaeology to Delight and Instruct: Active Learning in the University Classroom*, ed. H. Burke and C. Smith, 80–101. Walnut Creek: Left Coast Press.

Spencer-Wood, S.M., and M.C. Trunzo. 2022. Introduction to Archaeologies Special Issue on Intersectionality Theory and Research in Historical Archaeology. *Archaeologies* 18 (1) Special Issue: Intersectionality Theory and Research in Historical Archaeology: 1–44.

Stone, P. 2006. 'All Smoke and Mirrors…' The World Archaeological Congress 1986–2004. In *A Future for Archaeology: The Past in the Present*, ed. R. Layton, S. Shennan, and P. Stone, 53–64. London: UCL Press.

Sturdy Colls, C. 2015. *Holocaust Archaeology: Approaches and Future Directions*. New York: Springer.

Tarlow, S. 2001. Decoding Ethics. *Public Archaeology* 1: 245–251.

Ucko, P.J. 1987. *Academic Freedom and Apartheid: The Story of the World Archaeological Congress*. London: Duckworth.

Willey, G.R., and P. Phillips. 1958. *Method and Theory in Archaeology*. Chicago: University of Chicago Press.

Witmore, C., and M. Shanks. 2013. Archaeology: An ecology of Practices. In *Archaeology in the Making: Conversations Through a Discipline*, ed. W.L. Rathje, M. Shanks, and C. Witmore, 380–398. London: Routledge.

Zimmerman, L. 1995. Regaining Our Nerve: Ethics, Values and Transforming Archaeology. In *Ethics in American Archaeology: Challenges for the 1990*, ed. M. Lynott and A. Wylie, 64–67. Washington, DC: Society for American Archaeology.

———. 2006. Liberating Archaeology, Liberation Archaeologies, and WAC. *Archaeologies* 2 (1): 85–95.

Index

J. Carman, *Archaeological Practice as Politics and Ethics*,
https://doi.org/10.1007/978-3-031-69828-6

The manufacturer's authorised representative in the EU is Springer Nature Customer Service Centre GmbH, Europaplatz 3, 69115 Heidelberg, Germany. If you have any concerns regarding our products, please contact ProductSafety@springernature.com

Printed and bound by CPI Group (UK) Ltd, Croydon, CR0 4YY

30/09/2026

02225237-0002